T0127690

WATERSHED

WATERSHED

Herman Murrah

and the Pascagoula River Swamp

Davy Murrah

University Press of Mississippi / Jackson

Publication of this book is supported in part by funds
from the Mississippi Natural Heritage Publishing Initiative.

The University Press of Mississippi is the scholarly publishing agency of
the Mississippi Institutions of Higher Learning: Alcorn State University,
Delta State University, Jackson State University, Mississippi State University,
Mississippi University for Women, Mississippi Valley State University,
University of Mississippi, and University of Southern Mississippi.

www.upress.state.ms.us

The University Press of Mississippi is a member
of the Association of University Presses.

Manufactured in the United States of America
∞

Library of Congress Cataloging-in-Publication Data

Names: Murrah, Davy, author.
Title: Watershed : Herman Murrah and the Pascagoula River Swamp / Davy
Murrah.
Description: Jackson : University Press of Mississippi, 2024.
Identifiers: LCCN 2023046457 (print) | LCCN 2023046458 (ebook) | ISBN
9781496851949 (hardback) | ISBN 9781496852700 (trade paperback) |
ISBN
9781496851932 (epub) | ISBN 9781496851925 (epub) | ISBN 9781496851918
(pdf) | ISBN 9781496851901 (pdf)3
Subjects: LCSH: Murrah, Herman, 1935–2002. | Natural
history—Mississippi—Pascagoula River. | Nature
conservation—Mississippi—Pascagoula River. | Nature
conservation—Mississippi—Pascagoula River Watershed. |
Conservationists—Mississippi. | Pascagoula River (Miss.) | Pascagoula
River Watershed (Miss.) | Pascagoula River (Miss.)—Environmental
conditions.
Classification: LCC F347.P26 M87 2024 (print) | LCC F347.P26 (ebook) |
DDC 333.7209762—dc23/eng/20231115
LC record available at https://lccn.loc.gov/2023046457
LC ebook record available at https://lccn.loc.gov/2023046458

British Library Cataloging-in-Publication Data available

CONTENTS

WATERSHED

INTRODUCTION

Herman Murrah (1935–2002) lived his entire life on the banks of the Pascagoula River and in the Pascagoula River Swamp in southeastern Mississippi. To understand Herman Murrah, one first has to understand the Pascagoula River and the Pascagoula River Swamp.

The Pascagoula River is the largest unobstructed river in the contiguous United States. What is meant by "unobstructed" is the river has no levees or dams and the river channel has not been dredged or deepened. The river has been left to rise and fall naturally with the seasons. Because of this lack of restraint, the river overflows annually into the adjoining swamp and often completely covers the swamp. This phenomenon makes the Pascagoula River one of the wildest and most natural rivers in North America.

The Pascagoula River Swamp is not really a swamp at all—at least not what is normally envisioned when the word "swamp" is used. Most people think of something akin to the Everglades when they hear the word "swamp." The Pascagoula River "Swamp" is actually a bottomland hardwood forest, but the word "swamp" rolls more easily off the tongue. The Pascagoula River Swamp is one of the most ecologically diverse tracts of woodlands in North America. It is home to an unbelievably large number of species of plants and animals, including several endangered species that can be found nowhere else on earth.

When discussing the Pascagoula River and the Pascagoula River Swamp, it is best to consider them to be one entity because,

Herman Murrah at home on the Pascagoula River.

in reality, they ARE one entity. Neither the river nor the swamp could exist in their present forms without the other. In fact, during the spring floods, it often becomes almost impossible to discern where one stops and the other starts.

To be historically correct, one must add Herman Murrah to the mix of symbiotic relationships, when it comes to the Pascagoula River and the Pascagoula River Swamp. The river and the swamp formed the roots and foundation of Herman Murrah, and he would not have been the same person without them. Conversely, the swamp and river would not exist in their current forms without the influence of Herman Murrah. He was nurtured and formed by the river and the swamp, and he spent his life nurturing them in return. Their relationship was both symbiotic and spiritual.

CHAPTER ONE

Herman's Family

It was a beautiful day in the summer of 2002, but Herman Murrah couldn't help but be a little depressed. He knew his time was near. He by no means feared death, for he was ready to meet his maker, but he just wasn't through living yet. This was to be his last ride, and he knew it. He was on the way to the hospital to die, and there was no getting around it. His body was failing him on several fronts, but his spirit was as strong as ever.

Herman had regrets, as all men do, but he had a lot to look back on with pride. In his short stint here on this earth, he had accomplished enough to be proud. He was leaving the place in better condition than it would have been had he not lived. That was and is all any of us can hope for. He had made a difference.

His oldest son was driving on this fateful trip and Herman asked him to drive slowly so he could have one last look at the swamp he loved so dearly. You see, Herman loved his wife and his two sons. He had also loved his job, and he was incredibly thankful that he did because he realized that few men could truthfully say that. To him, his career had been more of a calling than a job. Several opportunities had presented themselves over the years that would have enabled Herman to make a lot more money than he was making, but none of those opportunities held the allure of his chosen career. Herman had been blessed with the opportunity to make his living in the swamp and on the river. He felt a quiet satisfaction that he had given back to the swamp as surely as the swamp had given to him. Herman Murrah's lifelong love affair had been with the Pascagoula River Swamp.

Herman's love affair with the river and the swamp began as a child of Curtis (also known as Curt) and Laura Murrah on the bank of the Pascagoula River. I say "on the bank," but from time to time during the spring floods, they would find themselves literally surrounded and engulfed by the river. The Murrah household's life and living revolved around that river.

Curt Murrah was a bit of a rascal. He was short of stature, and his head hadn't had the luxury of hair for about as long as he could remember. Having had very little formal education, Curt was far from imposing in any physical or intellectual sort of way. He was, however, wise in the ways of the world and a determined sort of person. Nobody intimidated Curtis Andrew Murrah.

Curtis made his living off the river and the swamp. Working at a regular job just wasn't for him. He worked hard and he played hard, but he just couldn't bear to punch a clock.

Curtis Murrah was a master of building simple boats. The boats were made of cypress boards cut from logs he salvaged from the swamp or river or bought from some of his cronies. He built the boats one at a time, start to finish, in his yard. Once a boat was complete, he would slap a couple of coats of forest green paint on it, then start the next one. If you wanted him to build you a custom boat, that was alright with Curt as long as your idea of a custom boat happened to fit exactly into his idea of the perfect river/lake boat and you liked the color forest green. In other words, he built all the boats exactly the same size and configuration and even the same color. He was stubborn that way. With him, it was "my way or the highway." Maybe that was why he worked for himself instead of punching a clock for somebody else. Once a boat was complete, he would sell it for a set price. No negotiating. The boats he couldn't sell he kept for himself and rented them out to tourists and local fishermen.

Building boats was by no means Curt's only way to utilize the river for his living. He actually built the boats in his spare time. He also fished commercially, ran his own fishing guide business, killed alligators and sold the hides and meat, caught and sold fish bait, ran the Wilkerson Ferry from time to time, and sold roe (unfertilized fish eggs) for processing into caviar. He looked for any way he could find to make a living off the Pascagoula River or its adjoining swamp.

Curtis Murrah answered to no man. He lived by his own rules—not yours. One of his favorite sayings was, "Curt Murrah doesn't need anybody." Actually, it was a bit more vulgar the way he said it, but let's keep it clean and wholesome here.

Curtis Murrah was the kind of dad who intended to "make a man" out of his boys. He had three of them, with Herman being the youngest. Nothing he ever did would actually be called "child abuse" but he didn't dish out a lot of love either, with the exception of what most people would call "tough love." He expected nothing from anybody else but was demanding of his boys. It was part of their training.

Herman's mother, Laura, could not have been more different from Curtis. She was as humble and unassuming as any person who ever lived. She was about as close as anyone ever got to being a true Christian. She loved Jesus and was a fount of love to others. She doted on her children, particularly her baby, Herman. She was just stern enough to be a good mother but definitely preferred being kind and gentle. Contrary to Curt, she was loved and respected in the community. Laura Murrah was the antithesis of Curtis Andrew Murrah.

Laura kept the family together as a good wife and mother should. She also contributed to the family's livelihood by devising ways to create income based on their location on the riverbank.

The oldest son, Milton, was typical of a firstborn. He was independent, mature for his age, and dependable. The middle son, Carol, was typical of a middle child. He was a bit insecure and somewhat of a discipline problem. Both of Herman's brothers were quite a bit older than him, and his relationship with them both resembled more of an uncle/nephew relationship than that of brothers. There was also a sister—Glora Dean.

CHAPTER TWO

Early Formative Events

The Murrah family lived in a large house, but it was far from opulent. It had a flat roof that leaked from time to time. There was, of course, no air conditioning at that time, and the only heat in the house was from a chimney fireplace and a wood stove.

One day when Herman was a baby, Glora Dean was warming a blanket for him at the fireplace. Somehow, the blanket caught fire, and the fire quickly spread to her old-fashioned, loose-fitting dress. The next thing anyone knew, Glora Dean was running out into the yard with her dress on fire. To this day, nobody knows why she ran out into the yard but run she did. By the time her mother caught her, she was literally a torch. Glora Dean's mother threw her to the ground and rolled her over and over until the flames went out, then did all she could to revive her daughter, but the girl had apparently inhaled flames by that time, and she died in her mother's arms right there in the yard.

This tragic incident happened when Herman was too young to know anything about it or remember it. Because of Glora Dean's death, Laura doted on Herman even more than she probably otherwise would have. As Herman grew, this special treatment from his mother, coupled with the stern upbringing from his father, caused Herman to become especially close to his mother and somewhat distant from his father. These disparate relationships with his parents would affect Herman the rest of his life in ways both positive and negative.

By the time little Herman was approaching school age, his mother was spending a lot of time preparing him for school. She prepared him well. So well, in fact, that when Herman started school, he was bored. His boredom, along with his being spoiled by his mother, led to his having a lot of problems in school. Herman made good grades because he was so well prepared by his mother, but he was a terrible student.

Herman's early school years were a bit tumultuous. On top of being bored with his studies and spoiled by his mother, Herman was preoccupied with the river and swamp and didn't really have time for school. As far as Herman was concerned, he lived in a wonderland and school must be punishment for some unspecified sin. The fact that he was one of the smallest boys in his age group and had his father's sharp tongue didn't help, either. Herman found himself involved in a lot of fights. Surprisingly, he won his fair share of the fights because he was such an active, and therefore athletic, kid.

Herman wasn't involved in any extracurricular activities at school. He strictly rode the school bus to school and rode the bus back home in the afternoon. Other than school, his entire social life consisted of weekly church attendance with his mother and brothers and the inherent social interaction that comes with living on the river.

When Herman was six years old, the Japanese attacked Pearl Harbor. By that time, his oldest brother Milton had a family of his own, but it wasn't much later that the middle brother, Carol, joined up to fight in what came to be known as World War II. Carol's part in the war was destined to be what is commonly called "island hopping," which amounted to evicting the Japanese from a series of islands in the Pacific Ocean as part of America's effort to move ever closer to the Japanese Islands to enable land-based bombing runs on the Japanese homeland.

The enemy was both vicious and determined. The series of invasions in which Carol participated were each a special kind of hell. He saw and participated in wanton death, destruction, and atrocities that would haunt him and many others for the rest of their lives. The war changed Carol, and it wasn't a change for the better. Nowadays, he would have certainly been diagnosed with PTSD (post-traumatic stress disorder) and could have received some treatment, but back then, PTSD was called "battle fatigue," and soldiers were expected to just "get over it."

Getting over it was and is a lot harder than it sounds. Carol compensated by drinking. In fact, Carol quite literally became a "drunk." Alcohol ruled his life. He drove drunk, was involved in innumerable drunken brawls, and even married a woman with the same drinking issues, and they proceeded to fight seemingly day and night. Amazingly, the two of them stayed together for life, but it wasn't much of a life. Carol was predictably killed in a drunk driving accident many years later.

Ironically, Carol's struggles with alcohol had a positive effect on Herman. During Carol's war years, his mother fretted much like most mothers fretted knowing their sons were at war, probably engaged in battle, and stood a good chance of never returning home. These years were rough on Laura and it showed, but the subsequent years when Carol was drowning in alcohol were even worse for Carol and Herman's mother. During his formative years, Herman would lie awake at night and listen to his mother cry a while and pray a while over Carol. Those nights were hard on a maturing young boy. He loved his mother more than anything and couldn't bear to see her in such heartfelt misery. Herman blamed the alcohol and vowed that he would never put his mother or anyone else who cared for him through that living hell.

Consequently, Herman lived the rest of his life without ever taking up drinking. At some point in his late teenage years, he actually tried a beer once, and the taste was just plain nasty to him. The terrible taste, along with his vow not to become a drunk, resulted in that one beer being his only direct experience with alcohol. For the remainder of his life, Herman was what was back then called a "teetotaler."

CHAPTER THREE

Living Off the River

Growing up on the river and in the swamp with Curt and Laura was part "wonderland" and part "school of hard knocks." As soon as Herman was old enough, he started helping with the family's backwoods income structure. At first, he primarily helped his mother. He was too young to spend much time on the river with his father, and trapping and killing alligators was out of the question for a young child.

Herman helped his mother dig worms in the swamp for sale to the local fishermen and helped out with the household chores in their home, which doubled as the welcome center for the river visitors and the patrons of the Murrah enterprises.

One of Laura's specialties was making duck down pillows, comforters, and quilts. The Murrah family killed a lot of ducks back then and plucked each and every one of them. Most people abbreviated the duck-cleaning process by ripping out the duck breast that contains the majority of the meat in a duck and is a lot easier to clean than the whole duck. Not the Murrah family. No meat wasted here by taking shortcuts. The entire duck was plucked and all the meat was eaten, including the heart, liver, and gizzard. The remaining internal organs were then used for fish bait. Laura collected the down to use for stuffing her sleeping accessories. She was careful not to use any of the matured feathers along with their quills. Her creations were used by the family and sold to supplement their income.

Laura also cooked and served meals to the visitors of the river. At that time, the location of the Murrah household was a busy place. Their home was at the east landing of the Wilkerson Ferry, which was the only way to cross the Pascagoula River between Merrill and the Gulf Coast. Often travelers stopped at the Murrah house while waiting for the ferry to return from the other side of the river. This location was also a popular fishing and hunting landing on the river. Even though the Murrah home was a long way from the nearest town, there were plenty of visitors to the Murrah homestead.

Laura Murrah had a peculiar philosophy when it came to raising children. She was a strict believer in the Bible's teachings about raising children. She believed in "spare the rod, spoil the child" but applied that technique differently from most parents. She believed that a parent should NEVER strike a child in anger. Consequently, when Herman misbehaved and she determined that he needed a bit of corporal punishment, she made an appointment with him for the next day. She never forgot or changed her mind about one of these appointments. This was an effective methodology because Herman had twenty-four hours to think about what he had done and the impending punishment.

Little Herman's childhood was a lot more than unending chores and life lessons. He was a child and spent a lot of time playing in the woods and in the edge of the river. He didn't know it then, but he was learning an appreciation for the splendors of the Pascagoula River Swamp. Life in the swamp was a long line of discoveries. A better childhood would be hard to imagine.

As Herman matured, he started spending more time with his father and helping him with his river and swamp endeavors. This time period in Herman's life was when he really learned the swamp and all its intricacies. Curt Murrah was an excellent person to teach young Herman about the swamp. Few people knew the river and swamp better than Curt.

As mentioned earlier, Curt Murrah was a stubborn man. There was (and still is) a deep hole in the river at the McRae Bluffs just upriver from the Murrah homestead. Over the years, many people reported seeing an unusually big fish come to the surface of the water in that hole. One day, Curt saw the fish for himself. It was just a glimpse, but that was all anybody had ever seen of the fish. Curt decided he was going to catch that fish.

Curt stretched a parachute line all the way across the river with a couple of weights in order to keep it on the bottom and hung one weighted "staging" about ten feet long at the halfway point. He couldn't find a large enough hook locally that would fit his purposes, so he built one himself and tied it to the staging. He then cruised the riverbank with his .22 rifle until he could find the right-sized water snake with which to bait the hook.

Every afternoon, Curt would kill another snake to bait his single hook and every morning, he would check that hook to no avail. People were starting to make fun of Curt, including members of his own family. But it didn't matter. Curt just kept baiting that hook every afternoon and checking it the next morning. He was going to catch that fish!

And then it happened. One morning Curt checked the hook, and the fish he had been after for weeks was there. He tried running the line several times. Each time he got close to the staging, the fish would make a run and Curt would have to let it go. Finally, he decided he would need some help and went back downriver to the Murrah home for reinforcements. Curt and Herman took one boat and Milton and Carol took another boat and returned to try to get the fish.

Their plan was to simultaneously start from each side of the river and gingerly ease toward the center of the river and the big fish. The first few times they tried it, the fish bolted and they had to let him go. Eventually, they managed to get the fish to the top of the water and Curt shot him with his .22 rifle.

The fish was far too large to get into either of the boats, so they tied it to Curt's boat and towed it to the Wilkerson Ferry landing in front of the Murrah house. Here they attached a Jeep to the line and dragged the fish out of the river. They had no way to weigh him, so they just left him there for the day for all to see, including the people that had been making fun of Curt. The fish was an alligator gar, and the consensus opinion was that it weighed about 350 pounds. If this fish had been officially weighed, it almost certainly would have been the record for this species. Curt Murrah had caught his fish.

Herman realized during this period in his life that one shouldn't talk about either the river or swamp as separate entities. They were just different aspects of the same thing. The river wouldn't be the same without the swamp, and the swamp certainly wouldn't survive in its present form without the river. They were intertwined in a myriad of ways. At certain times of the year, it became impossible to determine where one stopped and the other one started. In fact, one didn't stop and the other one start. The river and the swamp were one and the same.

Herman started helping his dad with the wide range of tasks involved in making a living off the river and swamp. He learned how to make minnow nets out of screen wire and how to pick a good spot to hang the nets, and he became proficient at baiting the nets and retrieving the minnows for use or sale. Minnows could also be seined when the river was low.

Herman also learned when and how to rake crawfish for bait and when, where, and how to catch freshwater shrimp for bait. He also learned about gathering catalpa worms and river mussels. And there were always the wigglers, grubworms, and nightcrawlers that his mother gathered with her trusty worm digger. In a pinch, one could use "puppy dogs" (salamanders), chicken and other fowl entrails, and even soap under the right conditions. Bait was always a challenge, and the gathering of bait was often the hardest part of commercial fishing with hooks.

It was during these years that Herman learned how to be an effective fisherman. The cash cow of commercial fishing was catfish. For the rest of his life, Herman would point out the fallacy of people asking if the fish were biting. In Herman's way of looking at it, the fish were always biting. "They have to eat every day, so they're always biting," he would say. "You just have to put the right bait in front of them at the right time and place."

Herman learned how to catch catfish during rising or falling river levels, at high water levels and low water levels, hot weather or cold, and even when the river was flooded. Especially when the river was flooded. He learned how to fish on top of the water and on the bottom of the river, on top of the sandbars and in the deep bends, behind the sandbars, along the banks of the dead rivers and oxbow lakes, in the sloughs, and even out in the woods during floods.

Commercial fishing for catfish was, and remains, a full-time and year-round job. It is also hard work. In order to save money, one must catch or harvest his own bait, place and constantly adjust his hooks, bait the hooks at least once a day, and take the fish off the hooks at least once a day. Then the fish have to be stored or cleaned and eventually sold. Cleaning the fish helped with the profit margin. Cleaned and filleted catfish sold for considerably more per pound than fish "on the hoof."

Then there was net fishing. It was legal and productive to fish gill nets and hoop nets. There were also illegal slat boxes and wire baskets to be fished. Back then, there wasn't a lot of law enforcement, and Curt Murrah's fishing spanned a wide array of methods to bring fish to the market.

Gill net fishing was a favorite of Curt's. Fishing gill nets was a productive means to catch some of the biggest catfish. A gill net is simply a sheet of netting that hangs loosely in dead water. It is not baited in any way. The net catches the fish (often by the gills—hence the name) simply by the fish trying to swim through the net. Gill net fishing was also about the only way to catch the fish species that wouldn't bite a hook. Two of these non-hook-biting species were buffalo and spoonbill (also called paddlefish).

The buffalo were hard to sell but good to eat. The reason they were hard to sell was the difficulty involved in cleaning one of them. This particular fish has a lot of bones and a tough outer shell made of big scales. Cleaning them was part art, part science and a lot of hard work. Buffalo also didn't yield much in the way of fillets. The Murrah family ate most of the buffalo that Curtis caught.

Spoonbill was a different story altogether. Spoonbill are easy to clean, and they yield nice big fillets of firm meat. Part of the reason for the ease of cleaning and the resulting nice fillets is the fact that the spoonbill has no bones (much like a shark). There was also an added benefit to catching spoonbill. During the spring, the unfertilized eggs (roe) in the spoonbill can be processed and shipped off to plants for further processing into caviar. Roe for caviar is quite expensive and amounted to a bonanza for the Murrahs. None of the Murrahs had ever eaten any caviar, but their dinner table certainly benefited from the Murrah participation in this international and lucrative industry.

Herman and his dad fished the gill nets in the dead rivers and eddy sections of the river during floods and concentrated their gill net fishing during the spring when the spoonbill had roe. When the river was down in the summer, they would move their gill net operation to the land-locked oxbow lakes. They would restrict their summer gill net fishing to nighttime hours because the fish would ruin in the nets under the summer sun, particularly the spoonbill. The spoonbill fish dies pretty quickly when it can no longer swim freely (also a lot like sharks).

So Curt and Herman would put out their nets in an oxbow lake just before dark and run them periodically during the night in order to avoid losing any fish to the relentless summer heat. While the spoonbill was the most susceptible to dying and ruining in the nets, fortunately, they had a tendency to make a big splash on the top of the water when they became stranded by the net. Herman and his dad would hear the spoonbill get himself caught and go get him immediately. They would clean the spoonbill and put the meat on ice right there on the bank of the lake in the middle of the night. The operation made for a long and tiring night of fishing.

One strategy they would use that would hopefully increase their catch was to go to the end of the lake and beat on the top of the water with their boat paddles to try and get the fish to flee into the waiting net. This method worked to some degree. One night, Herman got into trouble with his dad when he took a shortcut by pouring some gasoline on the top of the water and setting it on fire. This panicked the fish so much, they literally tore up the net in their mad rush to get away. Herman never tried that again.

They would take the nets up at daylight in order to avoid trapping and killing fish that would ruin.

One of the more efficient and illegal ways to catch catfish was to "telephone" them. Since this method was considered to be the most illegal, telephoning operations were generally carried out under the cover of darkness. The best time of year to telephone fish was the summer when the river was at its lowest and the catfish were more concentrated.

This method utilized an old-fashioned phone with a crank on it. (Your cell phone won't work.) Cranking on these phones created a substantial electric charge. Two wires would be dropped to the bottom of the river. Then you would crank on the phone and "call" the catfish. The electric charge would stun the catfish and cause them to come to the top of the water in a bit of a stupor. This process didn't work on scaled fish, only catfish. Once the catfish were on top of the water in an addled state, all you had to do was scoop them up in a dip net before they recovered. This was a time-sensitive operation because the catfish didn't stay on top of the water for very long.

The Murrah clan and some of their friends developed a group strategy for telephoning catfish. They would get together several boats, each with two occupants/participants. Usually, this group effort consisted of five or six boats or teams.

The "lead" boat operated the telephone. One person would operate the outboard motor to position and keep the boat in the center of the deep hole that was the intended fishing grounds. The person in the front of the boat would lower the electrodes to the bottom of the river and crank the telephone to "call" the catfish.

When the catfish came to the top, they would be greeted by the "catch" boats. In each of the catch boats, one person would operate the outboard motor to position the "netter" in the front of the boat within reach of the addled catfish. This way, all the catfish that came to the top of the water could be netted before they recovered and returned to the depths of the river.

Once the telephoning and netting were done, the group retired to the riverbank and the sorting and dividing began. The intent was to divide the fish between the teams in a fair and equitable way.

One of the two-man teams would be designated as the dividers/sorters. It was their job to divide the total catch into equal piles for each boat/team. The divider/sorter team was also expected to build each pile with catfish of all different sizes. In short, it was their job to make each pile pretty much identical in quantity and quality.

Once the separate piles had been made, a lottery of some sort would be conducted to determine who had first choice of the piles of fish. In the decided order, each team would pick a pile. The sorter/divider team would get the last pile. This method of deciding who got what pile was an incentive for the divider/sorter team to make all the piles equal since they were to receive the last pile.

At the end of the operation, all teams generally had a pretty good pile of catfish for their efforts.

CHAPTER FOUR

Living Off the Swamp

The Murrah family spent a lot of time fishing and not nearly as much time and effort hunting. Selling game animals was both illegal (not that legality was much of a deterrent for Curt) and not very profitable. At that time in history, there were not too many deer in the swamp. They had been decimated by overhunting. And deer meat would have been the only game meat that would have been profitable if there had been enough deer to bother with.

This is not to say the Murrah family didn't do a fair amount of hunting. The Murrah dinner table often featured small game such as squirrel, rabbit, duck, and wild turkey. Since the Murrah family ate a lot of small game, Herman came to be a good hunter of it. Even in later years when the swamp came to be populated by deer (largely due to a deer relocation program in which Herman actively participated), Herman never did take up deer hunting in earnest.

When Herman was growing up, there were a lot of "wild" hogs in the swamp. Most of these hogs were actually tame hogs that were just "running wild." They actually belonged to somebody but were "free grazing" in the swamp. The males were often caught and cut (castrated) when they were piglets. These hogs were off-limits to anybody other than their owners. There were also quite a few hogs in the swamp that had not been domesticated for several generations. These hogs were truly wild, and personal ownership was a bit more nuanced. Then there were the "razorbacks" that inhabited the swamp. These hogs had been running wild for as long as anybody could remember. They were vicious and dangerous. Needless to say, the Murrah dinner table often featured wild pork.

The only animal in the swamp that could be hunted profitably was the alligator. Curt killed a lot of alligators, primarily in the summer when the river was low. The alligator provided both meat and skin that could be marketed. Curt didn't "fish" for them the way you now see it done on reality television. He simply cruised the river, lakes, and sloughs and shot them as they lay sunning on the bank. This task was both dangerous and hard work. Curt didn't trust this task to his sons until they were actually young men. By the time Herman was old enough to hunt alligators, the alligator population had gotten pretty thin. Curt had been an effective alligator hunter.

All of this hunting and fishing in the Pascagoula River Swamp served to teach a young Herman how to get around the swamp during both low water levels and flood conditions. Herman had an innate ability to travel the swamp without ever becoming lost. He always knew where he was and how to get to any place he wanted to be in the swamp. This ability to navigate within the swamp without concern for losing one's way is particularly impressive because, in the swamp, landmarks are hard to find and recognize. There are no mountains that can be seen in the distance. The swamp seems to engulf any intruder. To the untrained eye, it all looks the same. Many people have lost their way in the swamp, usually for a few hours, but sometimes for days. Herman was never lost for even a minute.

These were also the formative years when Herman began to understand and appreciate the swamp. The hardwood forest interspersed with the occasional cypress pond or oxbow lake seemed to go on forever. The incredible abundance of creatures and fauna in the swamp was being impressed upon Herman. The wonderment he learned during these years sustained him for the remainder of his life.

Herman definitely grew up a "country boy." However, when most people think of a "country boy," they picture life on a farm. Herman was certainly not a "farm boy." His version of country life featured Jeeps rather than John Deeres, boat trailers rather than cattle or horse trailers, wildlife rather than livestock, wild-caught fish rather than farm-raised catfish, the river and natural lakes rather than farm ponds, woods rather than fields, skinning pliers rather than corn shellers, mosquitoes rather than boll weevils, and a boat and motor as a tool rather than an occasional weekend getaway.

The skill set that Herman learned was also somewhat different from that of a farm boy. Herman didn't learn how to operate a tractor, but he could certainly operate a boat and motor with the best of them. He had no idea how to properly harvest corn or cotton but could skin a catfish or squirrel or harvest minnows from a homemade minnow net. He never milked a cow but learned how to harvest roe from a spoonbill. Herman wasn't very adept at mending fences but could mend a gill net or hoop net. He never participated in castrating a young bull or inseminating a cow but had lots of experience catching and castrating young wild pigs. Herman didn't learn how to plow a straight row but could back a boat trailer down a hill using a Jeep with no brakes. He never called the cows home from the back pasture in the late afternoon but learned how to call a gobbler on a spring morning.

When Herman was a boy, his dad bought him an early model BB gun. The BB gun was Herman's first real introduction to serious hunting. The Murrahs had fig trees and catalpa trees in the backyard. It was Herman's job to keep the cardinals out of the fig trees and the rain crows out of the catalpa trees. He killed a lot of cardinals and rain crows. To Curt Murrah, there was no shame in tasking Herman to keep the cardinals killed out of the fig trees and the rain crows killed out of the catalpa trees. The Murrahs ate the figs and made fig preserves and used the catalpa worms for fish bait.

Herman also used his skills with the BB gun to directly help supply the Murrah dinner table. Robins were regular fare at the table during the winter, and Herman even managed a few squirrels and rabbits along the way with his trusty BB gun. Since a BB gun is lethal at only extremely short range, hunting with a BB gun is excellent training for a young boy.

Herman also killed for close observation every bird species he could find. Remember, this was before the internet and even before cameras with telescopic lenses were widespread. Herman didn't have a long-range camera, but he DID have a BB gun and he had become an excellent shot with it. He learned a lot about the local and transitory bird populations this way.

Herman was not alone in his method of learning about birds. John James Audubon was one of the world's foremost naturalists and was the inspiration for what is now the Audubon Society. He first rose to fame for his bird illustrations. Did you know that his bird illustrations had so much detail partly because it was his practice to kill the birds for close observation and measurements in order to make his illustrations accurate? He didn't have the internet or long-range cameras, either.

In Herman's later life, he grew to be quite the conservationist and learned the folly of his ways, but as a child, he was rough on the local bird population.

CHAPTER FIVE

The Teenage Years and Becoming a Man

By the time Herman had reached his teenage years, he was an integral part of the Murrah family's livelihood. He was, in many ways, also a teenager, running around with his friends and chasing girls. Herman wasn't what you would call "popular" in that he didn't run with a particularly large crowd, and he certainly didn't run with the upper-class kids. Herman was a bit of a loner with a close-knit group of friends. He was from "the wrong side of the tracks," but that was alright with him—so were his friends.

Herman's best friend was a boy named Leon. Leon was more like a brother to Herman than his own brothers. They remained close for the rest of their lives.

Leon didn't live in the swamp but was no less of a backwoods country boy than Herman. Herman, Leon, and their little group of friends would go out some weekends to bars and "honky-tonks." These places weren't what would be called nightclubs. They were similar to nightclubs in that they were places to drink, dance, and pick up girls, but they were a bit rougher around the edges than a typical nightclub. Fights were so common that a night without a fight was a rarity.

Lots of fights were just fine with Leon. He actually enjoyed fighting. He would have a few drinks, then start looking for an opponent. To Leon, fighting was just another way to let off some steam and have fun. Herman wasn't particularly fond of fighting and didn't drink, so he became the de facto "designated driver" even though that particular term wasn't yet in vogue. Herman avoided the fights unless and until his friend Leon really needed some help. Herman's job was to pick up the pieces at the end of the night and get everyone home safe and sound.

Back then, the small town of Lucedale had two walk-in theaters, one drive-in theater, several pool halls, and a bowling alley, so drinking and fighting weren't all the nightlife available to a young man. Herman lived in a stable home located in a version of paradise and had a close group of friends to run around with. Life was good.

Life was so good that Herman determined he just didn't have time for school. He had never liked school anyway, and it was just time to move on. He quit school and started working with some of the logging crews in the Pascagoula River Swamp. Logging was no cakewalk, but Herman would rather be logging in the swamp than sitting in a classroom, sort of like the old saying, "A bad day fishing is better than a good day at work." As far as Herman was concerned, a day of hard labor in the swamp was better than an easy day in the classroom.

Over the next few years, Herman learned about logging. What trees were being taken, what the process was, and how much damage was being done. The Pascagoula Hardwood Company only allowed what is called "select cutting." The area to be logged would be "marked" by the timber company by applying paint on the larger trees that were the only ones to be cut. The process caused a lot of collateral damage, but the overall condition of the swamp wasn't hurt too badly since the medium-sized and smaller trees would be left to give the swamp a healthy head start on repairing itself. At least there were no large areas of clearcut.

There was a logging accident during this period of Herman's life that resulted in both of his ankles being broken. As a result of this accident, Herman wore boots for the remainder of his life to give his ankles the much-needed support.

During Herman's time with the logging crews, he eventually worked his way up the food chain until he was afforded the opportunity to operate some of the heavy machinery. The skills learned here proved to be helpful later in life when he needed part-time work during his Game Warden years and when he was managing the Pascagoula River Wildlife Management Area.

In his late teens, Herman met a young lady about his age named Gertrude Havard. He was instantly in love. The young lady was a waitress at one of the popular restaurants in town and lived in an apartment over the restaurant.

Gertrude was a local girl who was raised in a sheltered environment. Her dad was the domineering type living in a sea of estrogen. Gertrude had two sisters and no brothers. The Havards were country people, but they were "hill" people, too. To them, the swamp was a strange place populated by equally strange people. Herman had an uphill (pun intended) climb to ingratiate himself with Gertrude and her family.

Herman realized early on that he was going to have a hard time winning Gertrude's heart. All his early advances were shunned. Gertrude didn't want to associate herself with a "swamp rat." The term sounds derisive, and it was. The swamp was a mysterious and uninviting place to most people, and swamp people were generally looked upon in the community as strange and mysterious people. Curt Murrah's antisocial behavior didn't help any. But Herman was persistent, and eventually, he won Gertrude's heart.

Now, Gertrude's family was another matter. Mr. Havard didn't want his daughter to have anything to do with a swamp rat, and he made his feelings known. Herman and Gertrude kept seeing each other anyway. She worked and lived in town, so their seeing each other without having to deal with Gertrude's father was easy to pull off.

A few months later, when Herman and Gertrude decided to marry, they both knew there was no chance of having a traditional wedding with Gertrude's dad giving her away. So, they just went to the local justice of the peace and made the marriage official.

After the "wedding," Herman took Gertrude by her family's house to get her things. He stayed in the car. When Gertrude came out of the front door with her meager belongings, her dad stopped her. "You're not leaving here with him." Charlie wouldn't call Herman by name, and the word "him" was said with obvious contempt.

Herman got out of the car. Charlie was a big man, but Herman would fight for Gertrude if needed. Gertrude was obviously flummoxed. Herman told her if she didn't get in the car, he would never return. Then he waited to see if he had a bride and if he was going to have to fight for her. Gertrude got in the car, and Charlie didn't try to physically stop her. The crisis was past.

The newlywed couple moved into a small camp on stilts next door to Herman's parents, and Gertrude was introduced to life in the swamp. At first, she was overwhelmed. Everything about life in the swamp was so different from life on the hill. Gertrude had never been in a boat and certainly not on the Pascagoula River. She couldn't even swim. She was afraid of snakes and alligators and had no idea how to deal with the annual floods. Gertrude had a lot to learn about swamp living.

Curt Murrah didn't help any. He had been looked down upon all his life by these "hill people" and was in no mood for accommodation. Curt was his normal antisocial self. Gertrude wasn't sure about this guy.

Laura was a different matter. She welcomed Gertrude with open arms. Gertrude was, after all, the new bride of her precious baby Herman. She took Gertrude under her wing and proceeded to teach her all there was to know about swamp living. Laura truly loved everybody.

Since Herman now had a family to support, he felt it necessary to work at a "regular" job, so he took a full-time job at Fryfogle Hardware in town. It was good work and the Fryfogle family treated him well, but working in town just wasn't for Herman. After a short while, he started looking for a way to make a decent living in the swamp.

Herman found the next best thing to a swamp job. He landed a job with the Mississippi Game & Fish Commission as a Game Warden. His first assignment was as an assistant in the Red Creek Management Area in the next county. This wasn't a swamp job, but at least it got Herman back into the woods, and he figured he could eventually parlay the position into a situation where he could make his living primarily in the Pascagoula River Swamp.

Herman found it ironic that his new job required him to enforce the game and fish laws that he and his family had been violating on a regular basis for as long as he could remember. At least, since he was assigned to the Red Creek Management Area, he didn't have to immediately crack down on his own family.

Back then, a large portion of the Red Creek Management Area was a preserve where no hunting of any kind was ever allowed. There was even a tall fence around the area to prevent anyone from slipping in there and hunting. As you can imagine, there were a lot of deer in this area.

One day, Herman's boss received word that some biologists from Jackson would be down the next day, and they wanted to run a battery of tests and examinations on a freshly killed young buck. Herman's boss was tasked with providing the young buck and, naturally, he passed this task down to his young assistant.

The next morning, Herman showed up before daylight with his favorite rifle—a .22 Magnum with open sights. His boss was aghast! "We assured Jackson we would have a freshly killed buck today and you show up with a BB gun?!?!" Herman explained that he figured the biologists didn't want the young buck mutilated and assured his boss the biologists would have their specimen.

When the biologists arrived later that morning, they had a fine specimen of a young buck waiting for them with one single .22 caliber bullet hole behind his left ear.

As mentioned earlier, the deer population in the Pascagoula Swamp was quite low due to overhunting. It was about this time that the State of Mississippi started a deer restoration program. Deer were trapped using hog traps (the only difference between a deer trap and a hog trap is the deer trap has a top on it; otherwise, the deer will simply jump out of a hog trap) and transported to areas where the deer population was too low. This was Herman's first opportunity to pay something back to the swamp he so dearly loved.

Herman was a music lover. He was a big fan of this new thing called rock and roll, and he also was a fan of country and western music. The number one song for the year 1955 (the year Herman married Gertrude) was "The Ballad of Davy Crockett."

Nine months almost to the day after Herman and Gertrude were married, their first son was born in December. He was named Davy Herman Murrah after Davy Crockett. Now, "David" was a popular and common name, but "Davy" was generally considered to be a nickname rather than a proper name, so when Herman told the hospital his new son's name was Davy, they filled out the official birth certificate with the name "David." Herman had them line through "David" and change it to "Davy," and to this day, his oldest son's official birth certificate includes this modification.

The birth of Davy was hard on Gertrude. In fact, her surviving giving birth was in question at one point. After the birth, the doctor told Herman that Gertrude couldn't get pregnant for at least three years or it could cost her her life. Since the possible consequences of another pregnancy were so dire and birth control was not fully dependable, Herman took a vow of celibacy for the next three years.

One can only imagine how difficult and trying a vow of celibacy can be on a young man (and woman), but Herman was taking no chances on losing Gertrude. He threw himself into his work for the next three years.

Three years, eight months, and a week after Davy was born, the Murrahs welcomed their second son, Danny Wallace Murrah, into the world. The first Murrah child had been named after Davy Crockett, so Danny was named after Daniel Boone. Herman didn't much care for the name "Daniel," so he settled on Danny.

Based, at least in part, on Gertrude's difficulty giving birth, Herman and Gertrude decided that two children were enough, so surgical precautions were taken at this time to ensure no further pregnancies were possible.

This was a time of change in the Murrah household. About a year after the birth of Danny, Herman's mother, Laura, passed away after a long illness. Less than a year later, Curtis remarried and moved out of the swamp, so Herman and his family moved into the main house. It was also about this time that Herman was reassigned from the Red Creek Management Area to one of the two Game Warden positions in their home county of George. Herman could now make his living primarily in the swamp.

In a short period of time, the relationship between the Murrahs and the swamp had changed. No longer did the Murrah family reap the swamp's bounty for a living. Rather, the current Murrah family made their living off the swamp by protecting it. Herman liked this arrangement better. A lot better.

CHAPTER SIX

The Game Warden Years

Over the next few years, Herman was on a mission to give back to the river and swamp. Of particular interest to Herman was protecting the recently enhanced deer population in the swamp. After all, the reason the deer herd had gotten too small was overhunting and mismanagement. Herman was determined not to let that happen again.

One of the big issues with the deer population was "headlighting." Hunting deer at night puts the deer at a distinct disadvantage and can decimate a herd. Herman was determined to eradicate or at least minimize the headlighting. Consequently, Herman wasn't home a lot at night because he took this particular duty to heart. He wouldn't hesitate to wait in ambush for hours to catch a headlighter. The penalties were severe for headlighting, and Herman figured if he caught enough of them and the word got out, then the people would become a lot more reluctant to go headlighting. His well-earned reputation was actually sufficient to considerably reduce the frequency of headlighting.

Another issue that concerned Herman was the overfishing of the river. There were several ways to illegally catch catfish, and Herman believed that reducing the frequency of use of these illegal methods was necessary to preserve the catfish population in the river. These three methods were wire baskets, slat boxes, and telephoning. Wire baskets and slat boxes tended to catch smaller catfish in extremely large quantities, thus reducing the population of younger fish, which were the most prolific reproducers. Telephoning, if done right, almost completely decimates the catfish population in a specific area. Herman figured these methods were illegal for a reason, and he intended to put a stop to them or at least greatly reduce their occurrence. His ambushing tactics learned by catching headlighters worked the same way for telephoners. Since almost all telephoning takes place at night and in remote sections of the river, Herman would actually camp on the riverbank and wait for the telephoners. Since telephoning, like headlighting, carries heavy penalties, Herman only had to catch a few for his reputation to take over and cause a drastic reduction in the frequency of telephoning.

Herman was determined to preserve and improve the game and fish populations in the swamp and river, so he worked long and hard at his new missions, achieving limited results.

The method of enforcing the hunting and fishing laws was writing tickets. Herman wrote a lot of tickets. The local hunters and fishers weren't used to law enforcement being much of a deterrent, so a period of training was inevitable.

However, Herman wasn't fond of writing tickets on technicalities. Back then, the game wardens were evaluated and judged on how many tickets they wrote over the course of the year. Herman never topped the list. In fact, he was always near the bottom of the list. You see, there were several ways a game warden could inflate his tickets written, and Herman just didn't play that game.

One of the ways to write a lot of tickets was to catch somebody with over the limit of bream. If you are fortunate enough to get on a bream bed at the right time of the year, it is easy to catch over the legal limit. Each fish over the legal limit could technically be written as a separate violation. For instance, the legal limit back then was fifty bream, and if you caught seventy-five bream, then twenty-five tickets could technically be written. Some game wardens actually followed this practice and thusly inflated their ticket-writing totals for the year. Herman considered this practice to be an abuse of power and, therefore, would write only one ticket in this circumstance.

Another way for a game warden to inflate his ticket-writing totals was to write tickets for fishing without a license. Fishing without a license was, of course, a violation of the law and, therefore, a ticket could be written. There was, again, a technicality that could be abused. Nowadays, when you buy a hunting or fishing license, it expires one year from the purchase date, but back then, everybody's fishing license expired at the end of June. Since the Fourth of July weekend occurred right after the fishing license expiration date, that was a good time to write a lot of tickets. Herman didn't play that game, either. He figured a person should have a few days' grace period instead of playing "gotcha" with them.

In the winter, there was another way to inflate one's tickets written for the year. There were (and still are) legal shooting hours for ducks. These official shooting hours are based on the official sunrise and sunset for the day and for your location. In other words, the shooting hours changed every day and varied depending on your location. The opportunity to catch a duck hunter who was a few minutes outside of these ever-changing shooting hours was another way to inflate one's ticket-writing numbers. Herman knew that his portion of South Mississippi was in a "dead zone" by not being in one of the major flyways for ducks and, therefore, duck hunters in the Pascagoula River Swamp were already at a disadvantage to the duck hunters in the Mississippi River Valley or on the Eastern Seaboard, so he figured the local duck hunters deserved a break. Herman never wrote a ticket for shooting ducks outside of the official shooting hours.

One way Herman would hold an individual to the letter of the law was if said person was a member of Herman's family. Herman was careful to avoid showing favoritism to his family. That would be completely unacceptable in Herman's mind.

At one point, Curtis (Herman's dad) had gotten himself elected to the office of justice of the peace, which was essentially the most local of judgeships. Milton (Herman's older brother) was a constable at that time, and Herman was a Game Warden. The other brother (Carol) was the only member of the family who wasn't in law enforcement. One day, Herman caught Carol hunting rabbits out of season. Carol had to go before the local JP (Carol and Herman's dad) and was summarily convicted by Curtis. It was hard being a lawbreaker in a family of law enforcement.

In the winter of 1960 and spring of 1961 came the biggest flood the area had seen in a generation. The Pascagoula River hit flood stage before Christmas and pretty much stayed above flood stage until after Easter. Measurements differed, but the highest crest during this period was around thirty-one feet.

For comparison, the low reading in the summer is usually about three feet, the river starts to flow into the sloughs in George County at around eighteen feet, and back then, the river started flowing over the road to the Murrah house at around twenty-one feet. In other words, the river level was about ten feet deep at the deepest point on the road to the Murrah house.

It is hard for most people to even imagine the Pascagoula River flooded to that level and particularly for that length of time. Since that flood, the only time the river level approached that height was in the winter/spring of 1973–74, and it wasn't up for nearly as long as it was in 1960–61.

The new river crossing of Highway 26 was completed in 1950, and this was the first real river flooding test the new highway had seen. The highway crossed the Pascagoula River and the swamp about halfway between the Murrah home and Merrill. Merrill is where the Pascagoula River is created by the confluence of the Leaf River and the Chickasawhay River.

A tremendous amount of dirt had been brought in to elevate the highway as it crossed the Pascagoula River Swamp. Once the river reaches the levels it reached in 1960–61, the river flow is actually the entire width of the swamp, and this raised highway formed a dam of sorts of the flood-expanded river. The only openings in this "dam" were the bridge over the river and one more small opening to the east. The river was "piling up" on the upstream side, causing unplanned pressure on the roadway and bridge. When you drove across the swamp on the raised Highway 26, you could easily see that the water level was considerably higher on the upstream side.

This situation led to a concern among the engineers and officials that the entire span of highway could suddenly collapse, resulting in a catastrophe. Reportedly, the highway department was on the verge of blowing a hole in the raised highway to relieve some of the pressure but never did it because the river finally started to recede.

Gertrude wasn't about to stay behind the flood waters. She packed up the kids and moved out to her parents' home before Christmas and didn't return until after Easter. Herman stayed behind. He wasn't about to stay with the Havards, and, in any case, Herman wasn't going to abandon his home.

All the vehicles had to be moved out of the flood zone because, back then, there were no hills at the Murrah home that were high enough to keep the vehicles out of the flood. Herman moved his truck to the other side of the river, where the bluff was high enough to remain dry. Whenever Herman needed to go somewhere other than the flooded swamp, he would motor across to his truck and leave the boat tied up while he was away. In a sense, Herman had reactivated the old Wilkerson Ferry route since the east end of the ferry was where Herman's house was, and the west end of the ferry was where Herman was now parking his truck.

Herman stayed in their house throughout the flood. The river water actually flowed through the house for a time. Not under the house—through the house. Herman raised certain furniture and appliances above the water level by placing them on cinder blocks. For a while, he was sleeping at night in his bed while the river flowed through the house and under his bed.

Life on the Pascagoula River can sure get interesting at times.

During the succeeding years, Herman concentrated as much as he could on restoring and protecting the river, the swamp, and their inhabitants. The children were still small and mostly the purview of Gertrude.

However, being a game warden was not as simple as it would seem. When one is a game warden, one is also what is called a law enforcement officer (LEO), and sometimes situations arise that require an "all hands on deck" approach to law enforcement.

For instance, Herman was often called upon to assist in rescue and recovery operations when someone drowned in the river. Herman designed and built a drag for the express purpose of finding a body in the river. Since he lived on the river, knew it better than most, and was a LEO with a boat, he became the "go-to guy" when there was a drowning.

The duties of a game warden were wide ranging. For instance, the state decided at one time that there were too many beavers in the state, so they instituted a "bounty" on beavers. The State would pay you five dollars for each beaver tail turned in to the game warden. During that time, there was a constant flow of people through Herman's house, turning in beaver tails. Late one afternoon, one man brought several beaver tails and a live baby beaver. He said he just couldn't bear killing that baby beaver. Herman paid him for the tails and an extra five dollars for the baby beaver. The Murrah family tried to make the baby beaver comfortable on some blankets for the night, but nobody got any sleep that night. You see, a baby beaver's cries sound just like an infant human. The next day, Herman took the baby to a beaver pond and turned him loose in the hope that he would survive.

A few years later, the state decided the best way to reduce the beaver population would be to increase the alligator population.

So, one day Herman came home to quite a few corn sacks full of live baby alligators with instructions to disperse them throughout the swamp. Herman dutifully carried out this task. Sometimes it's funny how life comes full circle. Herman's dad had pretty much eradicated alligators in the swamp, and now he was restocking the swamp with alligators. Now there is an overabundance of alligators in the swamp, so the state has opened up a legal alligator season.

At some point in the early sixties, a man was running from the law with a little girl as hostage. The pursuit culminated in a standoff at Pine Grove Cemetery in northern Jackson County alongside Highway 57. The call went out over the radio for backup, and a wide range of local law enforcement arrived on the scene and surrounded the armed man and his hostage.

The standoff continued for quite a while until the man put a gun to the head of the little girl and shouted that he was going to kill her. At this point, it looked like this situation could turn out horribly.

A shot rang out from the woods behind the cemetery. The man fell to the ground with a .22 caliber bullet hole behind his left ear, and the standoff was over. The little girl was safe.

A sniper had slipped through the woods and positioned himself behind the man and the little girl. The gunman didn't know the sniper was there. The sniper turned out to be a young Game Warden from George County named Herman Murrah.

Herman was also involved in the occasional drug interdiction operation. Whenever local law enforcement set up a drug task force to take down a marijuana-growing operation, he always seemed to be involved.

In June 1966, Herman was called to Jackson. This was truly one of those "all hands on deck" situations for LEOs. The "March Against Fear," which started in Memphis, Tennessee, was to arrive in Jackson. Early in the march, James Meredith had been shot and the racial tensions only increased as the march approached Jackson. Martin Luther King Jr. had joined the march for the last leg or two, and the state government was extremely concerned that the culmination of the march would turn violent. The statewide assemblage of LEOs was tasked with stopping any and all violence from either side.

Herman's job was that of a sniper. He was positioned on top of one of the office buildings facing the expected rally site. He was provided with a high-powered rifle with a scope. The instructions given to Herman were to keep a close eye on King and the people nearest him. If any of them started anything that could be interpreted as inciting violence, Herman was instructed to take King out. So Herman spent several hours watching Martin Luther King Jr. through crosshairs. Fortunately, the situation never developed that would have required Herman to squeeze that trigger. To say this was a stressful day would certainly be an understatement.

This time period was also when Herman joined the Civil Air Patrol (CAP). The CAP would pay for your training and provide a plane for continuing training flights. This was a way Herman could finally get in the air. He had always wanted to be a pilot and could see no chance of ever doing it on his own—not on game warden pay. Herman completed the training and became quite proficient as a pilot. In fact, he was a bit of a daredevil. He could be seen skimming the river, performing aerobatics, and was even known to fly his plane under the Highway 26 bridge over the Pascagoula River from time to time.

Herman really enjoyed "pushing the envelope." He not only pushed the airplane's capability envelope, he pushed the envelope when it came to where and when he was allowed to fly. One day, he was flying over the Florida Panhandle in an area where he knew he wasn't supposed to be when he was intercepted by two fighter jets. They forced him to land at one of the Air Force bases on the Panhandle, and Herman was almost arrested. They held him and questioned him for several hours before they finally let him go, but their rerouting and delaying caused Herman to run out of fuel on his way back home and to arrive after the airport had closed.

He had to literally land a "dead aircraft" on an unlit runway at his deserted home airport late at night. If he hadn't made it to the runway, he probably would not have survived the crash.

Herman enjoyed taking people for plane rides and scaring them with his aerobatics. His old friend Leon was a favorite passenger/victim. For Leon, a plane ride with Herman was an exaggerated version of a carnival ride. These were exciting days for Herman.

Herman was involved in a serious automobile accident during this time period. Someone pulled out in front of him (when he was driving too fast, which was most of the time), and he had to leave the road. The car that caused the accident left the scene and didn't notify anybody. Herman was found a couple of hours later when he crawled back to the highway. He had several broken ribs and a punctured lung, along with numerous less severe injuries. A few days in the hospital ensued.

Unfortunately, game warden pay was meager. Herman had to supplement the family's income by working odd jobs on his days off and whenever he took some vacation time. He tried to spend most of this time operating some sort of heavy equipment such as bulldozers, backhoes, and tractors. His prior jobs with the logging crews had provided him with the necessary experience to become a generic heavy equipment operator. All this additional experience with heavy equipment would come in handy in later years when he was appointed Area Manager for the George County portion of the Pascagoula River Wildlife Management Area.

Herman's youngest son started school in 1965. With the house now childless during the day, Gertrude took a job delivering newspapers for the *Mobile Press-Register*. This was a seven-day-a-week job, but it provided additional income for the Murrah family. The job also provided the Murrah family with their first new car—a 1966 Ford Falcon.

Gertrude's first day on the paper route was a harrowing experience. She had ridden the route the day before with the gentleman from whom she was taking over, and she was nervous about her first day. She got turned around several times, had to do some backtracking, and was several hours late getting home. These were the days before cell phones and there were no pay phones on Gertrude's rural paper route, so she couldn't check in with Herman. He was despondent. He was out searching the county roads since he didn't know the paper route. He even stopped at each bridge he came upon and peered over the side. Herman had what would probably be diagnosed as a mild nervous breakdown that day and never fully recovered. He dealt with what the old folks called "a case of the nerves" for the remainder of his life.

A lot of Gertrude's family lived near a small town in Alabama called Brewton. The large family held an annual family reunion in the summer. Since Gertrude now had a reliable car for the first time, she decided to attend one of these family reunions. Back then, Brewton was a four-hour drive from the Murrah home, so Gertrude planned to stay overnight with some of the extended family, making the trip a two-day affair. Until her death, Gertrude maintained that she told Herman she and the two kids were making this trip, but either she forgot to tell Herman or he forgot that she told him. In any case, Herman lost his family for the weekend. This episode added to Herman's developing "nerve" problem.

The paper route was rough on a car. All that stopping and starting. By the time Gertrude's "new" car was a year old, it was pretty much worn out. One Sunday morning, the brakes failed on her and she crashed into a bluff. The oldest son, Davy, was in the front seat (unbuckled, as was most often the case in those days), and he crashed into the windshield. The car was totaled, and Davy needed thirteen stitches in his right temple. Gertrude was bruised and battered a bit but no serious injuries. The younger son, Danny, was asleep in the back seat and sustained no injuries.

Gertrude continued running the paper route for many years. Because of the paper route, the Murrahs had a relatively new car most of the time since a car would only last a couple of years on the paper route.

During these formative years for her children, Gertrude made sure they were in church regularly. Herman tried going to church but found that when he was there, certain people would take advantage of his absence from the woods. Herman actually noticed particular men who would leave the church when he arrived. He knew these men and had no doubt as to why they left church. They were headed to the woods with the certainty that no game warden would be an impediment to their plans. Consequently, the swamp became Herman's temple. He even had a couple of particular places in the swamp where he would go to talk with God and meditate.

When it came to raising children, Herman was determined to be a better parent than his dad had been. Curt had been a hard and demanding man. He didn't hand out much love and support. Herman was going to be more loving and supportive—more like his mom had been. He spent as much time as he could spare with the kids. Herman was also reluctant to discipline the kids. He believed in leading by example.

CHAPTER SEVEN

Boys and Their Toys

It has been said that "the difference between men and boys is the size and price of their toys." Herman epitomized that saying. Despite boats and motors being tools of the Murrah trade, Herman was always enthralled with the boat and motor. He was genuinely interested in any and all modes of transportation, both on the river and in the swamp. From an early age and for the rest of his life, Herman always owned at least one boat and motor, as well as a Jeep. These were the days before four-wheelers, side-by-sides, and Gators, so Jeeps were the preferred mode of transportation on the old logging roads in the swamp.

By early adulthood, one of Herman's favorite toys was an old Willy's Jeep. The Jeep even had a nickname—the "Green Lizard." Herman and the Green Lizard were inseparable for several decades. The Green Lizard had no roof, no windshield, and usually no brakes. When you operate a Jeep in the swamp mud as much as Herman did, it is virtually impossible to keep working brakes, so eventually, he just gave up on having brakes.

A wide array of children as well as adults were treated to a ride on the Green Lizard over the decades. Herman had mounted a winch on the front bumper, and there was almost nowhere you couldn't take the Green Lizard.

Herman had a sense of adventure and certainly had a sense of humor. It wasn't unusual for someone to challenge him and the Green Lizard by offering to follow Herman anywhere he wanted to go (kind of like the child's game of "follow the leader") without getting stuck and having to use a winch. What most people didn't understand was that Herman had a couple of advantages. For one thing, Herman had better tires than most of them. But his biggest advantage was his intimate knowledge of the swamp right down to each and every mudhole. Very seldom was anybody able to follow Herman and the Green Lizard without Herman having to pull them out of a bog at some point in the day.

As for Herman's (sometimes twisted) sense of humor . . . . One day, a guy by the name of Elwood (a longtime friend of Herman's) wanted to go on an extended Jeep ride with him. Herman almost never turned down an opportunity to put the Green Lizard through its paces, so the ride was on. A little while into the ride, Elwood realized he had left his chewing tobacco and asked Herman to go back so he could get it. Herman just offered Elwood a pack of cigarettes and said something to the effect of "these are made of tobacco. Just chew on them." Elwood chewed on cigarettes for several hours of Jeep riding that day, complaining profusely the entire time about how nasty cigarettes were to chew. As soon as they arrived back at the Murrah house, Herman pulled a pack of Red Man chewing tobacco from his back pocket and took a chew. Elwood came unglued. Herman had a bit of a twisted sense of humor.

There was, however, one limitation to the Green Lizard. It couldn't go in water over the hood. At certain times of the year, this was often a limiting factor that mattered. Well, Herman solved that problem in two ways.

First, Herman got his hands on an Army surplus Jeep that could be modified to run completely submerged. The whole electrical system was thoroughly sealed, and both the carburetor intake and tailpipe had adapters to extend upward as far as needed.

Herman modified this particular Jeep to ford the floodwaters on the county road going to the Murrah house. At times, Herman could be seen coming down the road in his Jeep. And when I say Herman could be seen, I mean "only" Herman and a couple of vertical pipes could be seen. The Jeep would be completely under the water. Herman would be standing on the seat and reaching below the water level to steer the Jeep.

In later years, Herman found another way to get around the Green Lizard's water depth limitation. Enter the amphibious vehicle, otherwise called the Hustler, and later, the Swamp Fox. Both of these vehicles were six-wheeled, all-wheel drive, amphibious vehicles. If the water got too deep, the vehicle would simply float and the wheels turning underwater would provide propulsion. Herman even crossed the Pascagoula River a few times in the Hustler. For a number of years, Herman and his family, friends, extended family, and adopted family enjoyed the Hustler and Swamp Fox.

But the Green Lizard was always in the background . . . ready and willing for the next adventure.

Herman's infatuation with boats was lifelong. He started operating a boat and motor at an early age while helping his dad run trotlines or nets. By the time Herman was an early teen, he was operating his own boat and motor while assisting in the family fishing business. The boat and motor were necessary tools of the family's trade but could also be a lot of fun.

As Herman moved into his middle teens, it was inevitable that he would find a way to race with his friends. This led to learning how to "soup up" the motors and even modify the boats for racing. One trick up his sleeve was that the boats his dad built and that he raced were flat-bottomed, so they could go in extremely shallow water. Herman increased this built-in advantage by waxing the bottom of his boat, much like surfboarders wax the bottom of their surfboards or skiers wax the bottom of their skis. By having a waxed flat-bottom boat, Herman could cut corners around the sandbars in the river. Many people were heard to complain that racing against Herman wasn't fair because he didn't stay in the river. This was a bit of an exaggeration since Herman almost never actually left the river, but he did have a big advantage since he could run in water only a few inches deep if he kept his speed up and the boat planed.

At one point, Herman decided he was going to teach his kids and other neighborhood kids how to water ski, so he purchased a boat just for that purpose. The boat he bought was a fourteen-foot fiberglass boat rated for a 35-horsepower outboard motor, and he proceeded to equip the boat with a 50-horsepower motor.

This was also the time period that saw the development of the Buzzard Roost version of a surfboard. Herman called it a "surfboard," but it bore little resemblance to an actual surfboard. Herman trimmed one end of a 4' × 8' sheet of ¾" plywood into a semicircle, wet the sheet and forced an upward tilt to the front end, installed a harness on the bottom of the front portion with which to pull the board and its occupant (victim), installed another harness on the top of the front portion with an axe handle to hold onto and steer with, and, voilà, the Buzzard Roost Surfboard was born. When pulled at high speeds behind Herman's boat, the surfboard was even more popular than skis. All the kids in the neighborhood still have fond memories of riding Herman's surfboard.

One of the neighborhood teenagers was a bit pudgy and not overly athletic, so he was unable to get up on the skis. He would just bog along behind the boat. Herman took this as a personal challenge. He was determined to pull the boy up on those skis. Solution: replace the 50-horsepower motor with a 110-horsepower motor. So, Herman now had 110 horses mounted to the back of a boat rated for only thirty-five horses. He would pull that kid up on those skis now! And he did.

For a time, that boat was the fastest thing on the river. Nobody wanted to race Herman anymore. He did, however, finally get to put her to the test.

A lot of the duck hunters in the marshes near the coast were running up under ducks as they lifted off the water and picking them off like "sitting ducks." This method of duck hunting was highly illegal, but the local game wardens couldn't do anything with the perpetrators because the standard issue game warden boat was equipped with a 20-horsepower motor so the duck hunters would simply run away from them.

Enter Herman and his overpowered boat. A bunch of game wardens gathered in the Pascagoula River near its mouth, where it becomes a network of interconnecting bayous running through the marsh. When one of the game wardens started chasing one of the perpetrators, Herman would be called into action. Quite a few illegal duck hunters were surprised that day to see Herman pull up beside them at top speed and instruct them to pull over. One guy swears he was doing 85 mph when Herman pulled up alongside.

Herman's ownership of a ski boat turned out to be fortuitous for another reason. A couple of local fishermen caught a monster alligator gar in their gill net one winter night. The fish was so big they couldn't get him into their fishing boat. The State Wildlife Museum had put out a call for a really big fish to display and Herman had notified all the local fishermen, so these two gentlemen figured this fish might be a candidate. They went downriver and told Herman about the fish, so he took his ski boat back upriver and helped them retrieve the fish. That same fish is on display to this day in the Mississippi State Wildlife Museum.

Herman's infatuation with boats ran the gamut. At one point, he had a Gheeno, which is a cross between a jon boat and a canoe. He found it to be ideal for boating in the flooded backwaters of the Pascagoula River Swamp. Also, he built a homemade jet boat when the technology first became available. He took the foot off a 70-hp Johnson outboard motor and installed a jet foot. Now he could really travel in shallow waters.

The boat he really wanted was an airboat. That's one of those boats propelled by an airplane propeller mounted on the back. He just never got around to building one.

Boats can be a tool, but they can also be a lot of fun.

When the boys were a little older, the Murrah family took up horseback riding. As always, the kids were first. Danny had a Shetland pony named Kokomo, and Davy had a Welsh pony named Prince. Later, Herman bought a saddle horse named Buckshot. This was a time period when the Murrah family resembled a farm family to some extent.

It turned out that Prince was quite the athlete. Davy and Prince started competing in local horse shows, riding the barrel races, the arena race, and the poles. Prince proved to be adept, so they won a lot of trophies. Unfortunately, most of Davy's trophies were for second place. There was another boy in the county (Larry Joe Baxter) who had a faster Welsh pony and usually won first place. Davy and Prince would occasionally beat Larry Joe and his pony for first place, and every now and then, Larry Joe wouldn't show up. In that case, Davy and Prince would usually collect three first-place trophies, but second place was a lot more common than first.

During this period, a typical Saturday night would consist of the family (and a few friends) loading up the horse trailer with all the horses and attending a local horse show competition. There they would all watch Davy and Prince compete. They would also ride the family's horses around the horse show arena with all the other casual riders. At the end of the night, Herman would take all the horses back home to put them away while Gertrude and the boys would run the Sunday morning paper route.

Herman didn't have a personal truck with which to pull the horse trailer and using the state vehicle would have been inappropriate, so he pulled the horse trailer to the horse show with the Green Lizard. All the family and friends would pile onto the Green Lizard and off they'd go. It got to the point where the arrival of the "Swamp Rats" on the Green Lizard would be announced on the loudspeaker at the arena. The Buzzard Roost Swamp Rats made quite an entry.

Also during this time, the typical Sunday afternoon featured a swamp ride with Herman commanding the Green Lizard loaded with women and children. The young men and boys would follow along on their horses. It was both a family and community affair. Favorite destinations for these Sunday afternoon rides were Lower Rhymes Lake, Josephine Sandbar, Peachtree Ridge, and the Indian Mound.

This was also the time period when the Murrah family had essentially adopted another family in the community. The father of that particular family had died unexpectedly at an early age of a heart attack, leaving four fatherless kids. Herman became a surrogate father to these kids. All family activities included this adopted family. Only one of those four kids remains alive at this writing, and Brenda Havard McMillan still considers Herman to be their surrogate father.

It was common during the summer for the Murrah family, their adopted family, and many others in the community to spend the afternoons and most of the day Saturday swimming or water skiing in the Pascagoula River in front of the Murrah house.

It was a time of family, friends, fellowship, and fun.

After a few years, there was a transition from horses to motorcycles. Again, the boys came first. The early indication of this upcoming transition was a small Honda version of a minibike that both of the boys rode along with their friends.

Then came actual motorcycles for the boys. At first, used bikes were introduced, but new motorcycles shortly followed. A couple of other boys in the community already had motorcycles, and others soon followed. Soon the boys in the community with motorcycles numbered over a dozen, and sometimes it looked like a juvenile motorcycle gang had been established.

The swamp was a favorite place to ride. There were sloughs to cross and sometimes water to ford. The sandbars (particularly

Josephine Sandbar) had those natural drop-offs to ride on. And, of course, there were miles and miles of old logging roads to ride with built-in mudholes to wheelie through. Sometimes there were impromptu races and even, occasionally, an organized race.

Another favorite place for the unofficial "Buzzard Roost Motorcycle Gang" was the local gravel pit. The pit had natural obstacles and even natural obstacle courses. It was a great place to challenge yourself and hone your skills. Various forms of competition would break out on a typical Saturday or Sunday afternoon, including "follow the leader" and "motorcycle chase." "Motorcycle chase" involved throwing a stick at a fleeing victim—an impending injury if there ever was one. The gravel pit was even home to "The Big Hill," which was a tall hill that not everyone could climb. The road between the gravel pit and the emergency room was well traveled.

Eventually, Herman bought a (used) motorcycle of his own. He believed it was crucial to participate in activities with his boys. Herman became a regular member of the unofficial Buzzard Roost Motorcycle Gang.

As was usually the case with Herman, he souped up his motorcycle in a way that wasn't obvious. Herman's small motorcycle would outrun any of the others of a similar size and horsepower. One of the better riders in the neighborhood even took Herman's bike to a couple of motocross competitions and successfully competed against bikes that were supposedly more powerful.

However, boats and motors always remained in the picture and the Green Lizard forever lurked in the background.

CHAPTER EIGHT

New Owners of the Swamp

The 1970s was an eventful decade for Herman, his family, and the people of Mississippi.

The decade began with Herman working as a Game Warden living on leased property on the bank of the Pascagoula River. At that time, the Pascagoula Hardwood Company owned the Pascagoula Swamp in its entirety, including the land where Herman's family lived. There were about a dozen camps at the Wilkerson Ferry location, including one that Herman owned and another owned by Herman's brother, Milton. There were also a few houseboats scattered about. Herman's family, however, constituted the only permanent residents. The rest of the camps and houseboats were weekend getaways for people who could afford such a luxury. Most of the camp owners were from Laurel, Mississippi, but a few were local. Life on the river was busy on weekends but settled down to a hum most weeks.

All the camp owners and Herman leased their campsites/homesites from the Pascagoula Hardwood Company. It had bothered Herman for a long time that he didn't actually own his own home. He had tried to buy a few acres from the hardwood company, but they wouldn't even consider it. He was stuck.

Herman did, however, have a good relationship with the company. They paid him a stipend to look after their land, including the leased campsites. The company had ceased timber cutting operations a few years earlier.

The Pascagoula Hardwood Company was a privately held company owned primarily by several wealthy families in Laurel. The company's owners were contemplating the sale of the property and figured if they just let the timber grow, the land would appreciate in value for the eventual sale. The timber operations hadn't been profitable for a long time, anyway.

The company owned a large camp across the river from Herman, and part of his job was to keep an eye on their camp. Davy was even mowing and raking leaves in the yard of the "company camp" so that members of company management could enjoy the camp occasionally without doing any maintenance.

The Murrah family had settled into a rhythm. Both of the boys were in school and doing well. Gertrude was running her paper route every day. Herman had settled into his job as a Game Warden. The river was rising and falling in accordance with its natural rhythm. Life was both good and simple. No complications to speak of.

That was all about to change.

One spring day in the early seventies, a young man showed up at the "company" camp across the river from the Murrah family. He seemed to belong, so Herman didn't harass him. He was apparently alone except for a small dog, and Herman would notice them lounging on the bank of the river or swimming in the river in the afternoon.

Then Herman received a complaint from a local fisherwoman—there was a nude young man lying on the bank of the river across from the Murrah homestead. Herman had to go investigate and thus began the lifelong friendship between Herman Murrah and Graham Wisner.

Graham was, indeed, skinny-dipping when Herman arrived. Graham didn't understand what the problem was but graciously agreed to put on some clothes. He didn't want any trouble with this local yokel. This first meeting between these two very different men was a short one. It would take a little time for them to warm up to each other.

Graham was visiting the company camp during his spring break from an elite Northeastern liberal arts university. He was a member of one of the families that owned the Pascagoula Hardwood Company and was familiarizing himself with the Pascagoula River Swamp. Graham had never actually seen the swamp before. It had, up until now, just been a ledger entry on his family holdings spreadsheet.

Graham and Herman were, at first, equally unimpressed with each other. To Graham, Herman was just a local, redneck, ignorant, backward, last-century anachronism who would have to be tolerated when at the camp. To Herman, Graham was another spoiled elitist who looked down on him from his high perch, but he would have to be tolerated since his family literally owned the land on which Herman and his family lived. To make matters worse, Graham was a typical hippie of that era. His hair was long and unkempt, and he probably had never worked a day in his life. Neither of the two men foresaw a long-lasting friendship in the making at this point.

The two men independently decided to just put up with each other when Graham was at the camp. There would be no conflict here. Neither of them wanted any trouble with the authorities. Graham looked at Herman as a necessary evil because of his badge, and Herman considered Graham to also be a necessary evil due to his birthright.

The first one to make a conciliatory move was Graham.

Graham had been exploring the swamp on his own and was deeply impressed by its beauty as well as its mystery. He had a lot to learn about this newly found treasure that belonged to his family and could think of nobody better to help educate him about the intricacies of the vast swamp than someone who had lived here for his entire life. It was time to make a friend, so Graham swam over to the Murrah side of the river for a visit.

Graham and Herman had a long visit and they realized that, although no two men could be more different on the surface, they shared a love for and dedication to the swamp. Herman's love of the swamp stemmed from a lifetime of learning to appreciate the swamp, and Graham's newfound infatuation with the swamp stemmed from the discovery that he and his family actually owned what he had come to see as a wonderland.

Over the next year or so, Graham became good friends with Herman. He also became friends with Davy, who was several years younger than Graham and was severely lacking in knowledge of the ways of the world but was surprisingly well versed when it came to the intricacies of the Pascagoula River Swamp. Graham and Davy spent a lot of time together. Graham had an unquenchable thirst for knowledge of the swamp, so he was continuously asking questions.

One day, Graham confided in Davy that something was weighing heavily on his mind. The families that owned the Pascagoula Hardwood Company were in negotiations to sell the swamp to a company whose plan it was to clearcut the swamp and plant cottonwood trees in it. Graham had, by this time, fallen head over heels in love with the diversity of the swamp—the river, the lakes and sloughs, the trees, the wildlife, and the fauna. He knew that once this planned sale went through, the swamp in its current form would be lost forever. Davy was shocked at the prospect and admonished Graham that he needed to do something.

Davy certainly didn't know what could be done and neither did Graham, so Davy suggested Graham confide in Herman. In Davy's mind, his dad always seemed to know what to do, and this situation should be no different. These two young and idealistic men even dared to dream that maybe the state could buy the swamp and preserve it in perpetuity for the people of Mississippi. Davy told Graham his dad wasn't much into politics, but he should at least know whom to contact to explore the possibility.

Graham got with Herman and explained the situation to him. He also told Herman about his and Davy's pipe dream about the state buying the land. Herman was understandably skeptical about this scheme but was wholly on board with the concept of "nothing ventured, nothing gained."

A dream had been germinated, and now, it just had to be nurtured to fruition. A two-pronged approach was undertaken.

Herman ran the idea up the Mississippi Game and Fish Commission flagpole. The idea eventually landed on the desk and in the heart and soul of the director of the Commission, one Avery Wood. Avery was immediately sold on the idea and started working on ways he could make it happen.

Meanwhile, Graham traveled back to Washington and started making connections through some of his influential family members. They decided an outfit called The Nature Conservancy would be a useful vehicle to guide this surely complicated transaction through to its conclusion. Dave Morine, the second in command at The Nature Conservancy, got hold of the idea, and it immediately became a mission for him.

From this point on, Herman, Graham, Avery Wood (and his assistant Bill Quisenberry), Dave Morine, and Charles Deaton (their ally in the Mississippi Legislature) became the sales force for this lunatic idea. They all realized the whole pipe dream was just that—a pipe dream. The chances were slim that they would succeed, but the reward would certainly be worth the effort, and for different personal reasons, it had instantly become a mission for each individual in this unlikely alliance of disparate characters. In fact, they were all working feverishly toward a shared dream when, at this point in time, few of them had ever met or even heard of each other.

Herman's world was changing fast. It was looking like the Pascagoula Hardwood Company was about to divest itself of the swamp one way or the other. The prospect was terrifying to him. It had always bothered Herman that he didn't own the land under his house. He had been hesitant to modernize the house or spend any money whatsoever on improvements. The house was too big to move if he ever lost the lease on the property. He had tried several times over the years to convince the hardwood company to sell him a little plot under his house, but they would not consider selling even a fraction of an acre. Herman was in a pickle. And now that the Pascagoula Hardwood Company was about to sell the land to someone else, he feared his new landlords would terminate the lease and he would have to find a new home.

What Herman really wanted to do was build a new house or at least raise the elevation of the existing house. He didn't really mind the spring floods or the river cutting him off from civilization and causing him to have to come and go by boat. Herman was as comfortable in a boat as he was in a vehicle. What bothered Herman was the rare occasion when the river would rise high enough to literally run through the house.

The people who had camps or houseboats could simply retreat to their homes when the river flooded, but the riverbank was the only home the Murrahs had. Though the oddity of the river running through the house was fascinating to others and made for a good story, the real-life experience was not a walk in the park. One detail that made the situation even worse was that the camps were built on stilts above the flood levels and the houseboats could rise and fall with the river level, but the Murrah home wasn't high enough to escape flooding. Something needed to be done, but Herman's hands were tied as long as he didn't own the land.

And now the hardwood company was about to sell the land.

Herman had cultivated and maintained a détente of sorts with the upper management of the Pascagoula Hardwood Company, but he had no assurance or confidence he could do the same with whoever became the new owners. He was looking at the very real possibility of losing his home.

Then came the flood of 1973–74. This flood rivaled the flood of 1960–61 as far as maximum river level. Fortunately, this latest major flood didn't last nearly as long as the flood of 1960–61, but it was a reminder of the predicament the Murrahs faced. Once again, Herman found himself sleeping on a raised bed in his own home with the flooded river running through his house and under his bed. The Pascagoula River was stressing the point to Herman that something had to be done.

So, while the original group of dreamers was diligently trying to ensure the Pascagoula Swamp wasn't lost for all time, Herman had another mission of equal importance. He had to do something about the situation regarding his home, and he had to do it quickly. The status quo was about to change.

Herman realized that one thing had changed over the last couple of years—he had made a new friend in the Pascagoula Hardwood Company—Graham Wisner. When Herman mentioned his predicament to Graham, his concerns didn't fall on deaf ears. Graham was aghast at the thought that his extended family wouldn't come to Herman's rescue. This was, after all, Herman's home, and Herman had always been a friend to the Pascagoula Hardwood Company. And he was certainly Graham's friend. Graham allied with Herman in this new quest.

The three gentlemen who were essentially running the Pascagoula Hardwood Company were Bill Chisolm, Bob Hinson, and Gardiner Green. These three were the patriarchs of the three largest families in the hierarchy of the Pascagoula Hardwood Company. The other family was the Wisners. Herman's primary dealings up to this point had been with Bill Chisolm and Bob Hinson. Graham knew all of these people but not too well. He started working the issue from a family standpoint, and Herman started working the issue based on his longstanding relationship with Bill Chisolm and Bob Hinson.

At first, the idea of selling Herman some land was not well received. The hardwood company was already negotiating with a large corporation and didn't want to throw any kinks into the apparatus. But both Herman and Graham were insistent.

Graham's family was one of the four families that owned almost all of the Pascagoula Hardwood Company, but the Wisner family was the smallest of the families based on the percentage of stock owned. Graham was, of course, only one member of the Wisner family, so Graham was, in fact, an extremely small shareholder in the company. Also, the Wisners had long ago moved away from Laurel and had no part in running the company. Consequently, Graham's influence within the company was minuscule. But Graham was on a mission, and he wasn't going to leave his friend in a bind.

Graham had often fantasized about his small percentage of shares being converted into actual acreage and his owning that small portion of the swamp. Things didn't work that way, but it was a favorite way for Graham to imagine the situation. The Pascagoula Hardwood Company owned approximately 40,000 acres of the swamp, but Graham's theoretical portion of the swamp amounted to only a few acres. Graham started a campaign to have his shares converted to land with said land being given to Herman. This was never going to happen, but even the proposal of this "out of the box" scheme demonstrated how important it was to Graham that Herman be assured of a place to live in the swamp. Graham was NOT going to let his friend down.

After months of wrangling, pressuring, and just generally insisting on behalf of both Herman and Graham, Bill Chisolm, Bob Hynson, and Gardiner Green finally relented to the idea of selling Herman a homestead. The acreage, price, and location were agreed upon, so Herman brought in a local survey company and had the agreed parcel surveyed and a plat drawn up. In 1974, a deed was executed between the Pascagoula Hardwood Company and Herman Murrah. Herman finally owned the land upon which his house sat.

Unfortunately, Gardiner Green had, at the last minute, changed the agreed-upon deal and reduced the acreage as well as eliminated the riverfront aspect of the parcel. This change had been made at the last minute and without consulting with Herman. Herman didn't even realize the change had been made until after the "deed was done." (The final official deed recorded in the George County Courthouse records even shows those marked through and handwritten changes to the plat.) Herman was both elated that he now owned his own home and furious that he had been swindled in the dead of night.

The longshot plan to get the Pascagoula Hardwood Company lands sold to the State of Mississippi involved a lot of politics, negotiations, and financial wrangling. Most of these negotiations were going on behind closed doors in Jackson and Laurel.

Herman's part in all this at this point was to function as part diplomat and part salesman. He was one of the main emissaries for the swamp. It was his job to convince everyone that saving the swamp was not only important but crucial. Herman hosted several tours of the swamp for dignitaries, public relations people, film crews, politicians, and participants in the negotiations.

If the state was going to ante up such a large chunk of money, the public had to be brought on board with the plan. The Legislature was going to have to be pressured by their constituents or it would never happen. Consequently, a statewide public awareness campaign had to be put into place. Avery Wood placed a young man named Bill Quisenberry in charge of this campaign. "Quiz," as he became known, and Herman spent a lot of time together assembling the case that Quiz was going to present to the people of Mississippi. These two men remained friends for life.

In the meantime, Herman had a house to build. Since he now owned the land, he was going to do this right. The first priority was to build his new house above the flood level. The simple way to do that was to build it on stilts like all the fishing camps in the area. Herman didn't want a house on stilts. He looked ahead to the later years when he and Gertrude would struggle with a long flight of steps to their front door. He had another idea.

The Murrah home was going to be built on a hill. The only problem with that idea was there was no hill on the Murrah property. No problem—build the hill first, then build the house. So, Herman bought an old dump truck and started hauling dirt from the same dirt pit that had hosted so many motorcycle rides in years past. The dirt hauling went on for months until Herman was satisfied that he had built a big enough hill to stay above the floods. The Murrahs had marked several of the surrounding trees with the high water marks of the two big floods in 1960–61 and 1973–74, so Herman knew exactly how tall the hill had to be. Once all the dirt was in place, Herman let the dirt settle and pack for almost a year before he started building his new home.

Home financing was arranged, a contractor was hired, and Herman's house was built—to his specifications. During the construction process, Herman oversaw everything that was done. He used only top-of-the-line materials and ensured that no shortcuts were taken. This would be the only house Herman would ever build, and it was going to be done right. A few months later, the Murrahs moved into their new home. By now, Davy had married and moved to Starkville, where he was attending Mississippi State University, so Herman, Gertrude, and Danny were the first occupants of the new Murrah dwelling.

Herman finally had a home that was truly his. A lifelong dream had been realized.

Meanwhile, the longshot effort to have the state buy the swamp was ongoing. It was starting to look like this might actually happen. The "families" had, for the most part, been convinced to go through with the sale. The issue now was whether or not Avery Wood and Charles Deaton could convince the Mississippi Legislature to sign on. After all, without their approval of the funds, nothing could possibly move forward. "Quiz" was conducting a statewide full press campaign to garner public support. The plan was to have the public pressure the Legislature to approve the funds. Herman could only cross his fingers. He had done all he could do. But he worried that the deal could still easily fall through and his precious swamp would be lost forever.

There was also game warden work to do. The word had gotten out that the swamp might be destroyed, and certain people saw this as a last opportunity to take as much as possible from the swamp while it still existed. Herman wasn't going to stand by and hope for some kind of legislative salvation while the swamp was pillaged and raped. He had a job to do.

Gertrude continued running the paper route, and their two boys were nearing graduation—Davy from Mississippi State and Danny from George County High School. Life continued happening while the whole world (or at least South Mississippi) waited on the State Legislature to act.

How would things look in a few years? Would the boys finish school and go on to make their own lives? Would the state buy the swamp? Even if the state bought the swamp, what would happen from there? Would the Murrah family find themselves living in the middle of a giant cottonwood plantation without a squirrel, deer, turkey, or rabbit to hunt? What would the clearcutting of the swamp do to the river and the oxbow lakes? Was Herman destined to live in a paradise of the ecologically diverse swamp he had grown to love, or did his future consist of living in a cottonwood "desert"? So much was at stake, and Herman felt helpless to do anything about any of it. He checked in from time to time with his friends Graham and Quiz to see how things were going but could only sit on the sidelines and pray.

Herman worried about his future grandchildren and, for that matter, their grandchildren. Would they be afforded the opportunity to hear a wild turkey gobble on a spring morning in the swamp? Watch as a mother wood duck and her brood swam by? Relish in the pure joy of watching a couple of river otters playing in a slough? See a blue gill rise to a fly on a summer morning in

Lower Rhymes Lake? Watch an ivory-billed woodpecker in flight? Marvel at a rabbit sitting on a floating log in the flooded backwoods? Rake some crawfish from the backwater in order to bait some hooks in Crooked Slough? Or even see a swallowtail kite conduct aerobatic maneuvers over the Pascagoula River Swamp?

So much was at stake. How would it turn out? What could Herman do? Sadly, nothing but watch, listen, wait, and pray.

On May 2, 1976, HB914, as amended, was passed by the Mississippi Legislature. The bill authorized approximately $13 million and would eventually result in the state acquiring approximately 32,000 acres of the Pascagoula River Swamp. However, the deal was not yet done. The plan was for the Nature Conservancy to acquire 75 percent of the outstanding stock in the Pascagoula Hardwood Company. It was 75 percent instead of 100 percent because Gardiner Greene (the same guy who had swindled Herman by reducing the size of the Murrah property and taking away the riverfront) had worked out his own side deal. Gardiner Greene was going to end up with a big chunk of the land currently owned by the Pascagoula Hardwood Company. In addition to swindling Herman and now throwing a monkey wrench into the deal to save the swamp and almost causing the deal to fall through, Gardiner Greene's lack of honor would again rear its ugly head a little later.

Once the Nature Conservancy acquired 75 percent of the stock, they planned on liquidating the company, resulting in the Conservancy owning most of the land and Gardiner Greene owning the rest. At that point, the Conservancy would sell the land to the State of Mississippi. This sale, which was the ultimate culmination of the deal, took place on September 22, 1976.

(Years later, Gardiner Greene clearcut his portion, creating a vast wasteland, and then sold it to individuals. Mr. Greene was certainly within his rights, but Herman always regretted the original purchase didn't include all this additional land.)

The deal was done! The swamp was saved! The nightmare of an expansive cottonwood plantation had been avoided! The wildlife, flora, and fauna of the Pascagoula River Swamp would survive! Herman and his descendants would have the opportunity to live in paradise rather than an ecological "desert" of cottonwood trees! If Herman had been a drinking man, he would have opened a bottle of Champagne.

But what now? Questions abounded. How would the area be managed? Locally or under the heavy hand of Jackson? Who would be the area manager? The area spanned two counties. Would the counties be managed separately? What about the camps on Pascagoula Hardwood land that was now land belonging to the State of Mississippi? What about the houseboats moored to the riverbank now owned by the state? Was a survey to be done of the area? A fence installed? What improvements were to be made? What roads were to be improved and what roads were to be closed? What was to be done with the Big Swamp? (The Big Swamp is a large area between Black Creek and the Pascagoula River to which the state had no direct road access.) Would additional and improved boat ramps be installed? If so, where? Would food plots be planted? Where? How many? Would funding be made available to improve, maintain, and manage the area, or would it just sit fallow? What part would politics play? What rules would be developed for the area, and who would have input into their promulgation? What usage fees, if any, would be imposed? Questions truly abounded. And now these questions would have to be answered. But who was to answer these questions? That was the first question.

But Herman's main question was what would be his involvement? Would he actually be involved? After all, Herman was not a political man. He had no power and very little influence. What he DID have, however, was a passion for the swamp and the river. It was his fervent hope that he would be allowed to channel that passion into action.

So many questions to be answered.

The first question, "who was going to be calling the shots?" was answered immediately. That would be the politicians in Jackson.

The jockeying for position and power had actually begun well before the deal was done. This was, of course, no real surprise to anybody. Politics always raises its ugly head in a case like this.

The people who had worked so hard for so long to get this deal done were, for the most part, pushed aside by politics.

Avery Wood was no longer the director of the Mississippi Game & Fish Commission. His was an appointed position, and since the governor had changed, he had been replaced by this time. Avery was completely out of the picture.

Dave Morine and the Nature Conservancy had been a crucial cog in the machinery that brought the purchase to fruition, but their part in the Pascagoula River Wildlife Management Area was pretty much done. Dave and the Nature Conservancy would move onto other projects. The Conservancy would continue to be active in the State of Mississippi and remains so at this writing.

Charles Deaton remained a powerful member of the Mississippi Legislature, but his involvement in the Pascagoula River Wildlife Management Area would be pretty much limited to that of any other legislator.

Graham Wisner was no longer part owner in what he considered to be the wonderland of the Pascagoula River Swamp. Graham would go on to make a name for himself in international law and still lives on the outskirts of Washington, DC. Many members of his extended family still resent Graham for what they consider to be a bad deal that they made when they sold to the State of Mississippi. They believe that Graham's idealism cost them a small fortune.

Bill Quisenberry ("Quiz") remained with the Mississippi Game & Fish Commission. His good work ethic, his penchant for detail, and his natural humility endeared him well with the upper management of the Commission. Bill remained with the Commission until his retirement many years later. Quiz's position in the upper echelon of the Mississippi Game & Fish Commission proved to be an asset for the new area manager of the Pascagoula River Wildlife Management Area.

And that new area manager was none other than Herman Murrah. Actually, Herman was named Area Manager for the "Upper Pascagoula River Wildlife Management Area." In other words, he was over the George County portion. A gentleman by the name of Jim Kirkwood was named the area manager for the "Lower Pascagoula River Wildlife Management Area," or the Jackson County portion.

Herman was, indeed, going to be given the opportunity to put his passion for the river and swamp to work. And the real work was just about to begin.

CHAPTER NINE

Managing the Swamp

Most of the previously unanswered questions remained unanswered. It was determined by the powers in Jackson that an area headquarters was to be set up in each county. The George County Area Headquarters was to be built just south of Herman's house on the bank of the river. This was to be Herman's first large task—lay out and supervise the construction of the area headquarters.

As far as "managing" the area, there was a lot more that needed to be done other than building a headquarters. Herman soon found himself stretched in many directions at once.

As for the remaining unanswered questions about how the area was to be managed, the state set up a committee. Of course they did! Herman and the other area manager were on the committee. The committee members represented a wide array of interests. As is the case with a lot of government committees, the primary interest of many of the members was "self-interest." There was a forester, a biologist, a couple of "tree huggers," and numerous people who can be described no other way than political appointees. It was the task of the committee to come up with a proposal for a ten-year management plan. The proposal would, of course, be subjected to final approval by the powers in Jackson.

The meetings and discussions of the committee were private. They were also heated. Several members of the committee wanted the area to essentially become a preserve with very little to no public access. The forester saw an opportunity to "manage" one of the last great hardwood forests in the South. The biologist was all about the animals, particularly the deer herd. Several members just wanted to make a name for themselves.

Herman, of course, had a different view. He believed that the good people of the State of Mississippi had rallied in support of the state buying the property and had ponied up the funds to do so. While nobody in that committee room was more passionate about the swamp and its inhabitants than Herman, he became the primary voice for the people. Herman figured it did no good to have a preserve that was made inaccessible to the people who had sacrificed so much to save his precious swamp. He wanted to strike a balance—preserve the swamp FOR THE PEOPLE. He wanted better access for the people while most of the committee members saw the swamp as a preserve (the "tree huggers"), a source of funds (the forester), a deer farm (the biologist), or as political leverage (the political appointees). Most of the committee members saw the swamp as an opportunity to further their particular interests while Herman was looking out for the swamp, its inhabitants, and the interests of the people of the great State of Mississippi.

In the meantime, Herman had become a grandfather. His first grandson was born in Starkville (the home of Mississippi State University) and was named Joshua David Murrah. Joshua was born with pneumonia and it was touch and go for a couple of days, but he pulled through. Herman was now "Grandpa Herman." The birth of Herman's first grandchild only served to strengthen his resolve to preserve the swamp and make it accessible to the people so his grandchild(ren) could have the opportunity to appreciate and enjoy the wonder of God's creation the way Herman had.

Herman would remain a defender of the swamp and the people of Mississippi and a bulwark against the unrelenting forces of political self-interest for the remainder of his career. This calling was a natural extension of Herman's lifelong love affair with the swamp. Unfortunately, Herman would have to endure the wrath of the politicians (which would eventually cost him his job) as well as the criticism of numerous citizens.

Abraham Lincoln paraphrased the poet John Lydgate when he said, "You can please some of the people all the time and all the people some of the time, but you can't please all the people all the time." Situations were about to arise where Herman would find himself the reluctant enforcer of rulings with which he strongly disagreed and, consequently, the target of the ire of many of the very people whose interests he had fought so hard to preserve and defend.

One of the main issues that caused Herman a lot of heartache was the issue of roads into the swamp. The swamp was currently crisscrossed by a network of old logging roads. Most of these roads were impassable to the average person. In order to utilize them, one needed a four-wheel-drive vehicle. Herman knew that most people didn't have such a thing and would, therefore, not be able to access the bulk of the swamp. Something needed to change.

In Herman's mind, at least a few areas in the swamp needed to be made accessible to the average citizen who didn't possess a four-wheel drive. Many people disagreed with Herman on this concept. It was an odd assortment of people that opposed Herman when it came to installing some good roads into the swamp.

First, there were the "tree huggers" who really didn't want any improvements whatsoever. They envisioned a wilderness unblemished by the intrusion of man. This attitude was anathema to Herman's way of looking at things. What good was it to preserve the swamp if nobody was allowed to enjoy it?

Then there were some of the old-timers who had hunted in specific areas of the swamp for generations and were dead set against making those areas accessible to the general public. Some members of this group were the most vocal and participated in personal attacks against Herman, including many formal complaints that found their way to Jackson. Herman tried vigilantly to avoid taking these attacks personally, but in reality, his heart was broken that so many people failed to realize that he was doing his best to defend both the swamp and the people that deserved the opportunity to enjoy its infinite wonders.

Finally, a compromise was reached on the roads. Improved roads would be built to some of the more popular lakes and to the Josephine Sandbar. The remainder of the old logging roads would be gated off and made off-limits to vehicles. Herman was not happy with the compromise, but that's the nature of compromise. With most compromises, nobody is really happy because nobody gets exactly what they want. Herman would have to make it work.

Since Herman would be the face of the Mississippi Game & Fish Commission in George County, he also knew he would have to take the heat from the people who weren't happy with the compromise. Herman caught it from all sides—the people who didn't want any roads built, the people who didn't want any roads blocked off, the people who wanted more roads built than were planned and, of course, those from Jackson who thought the roads were costing too much. There were even some people who complained about the quality and design of the improved roads. Some thought the roads should be better (some even wanted them paved), while others complained that all the excessive roadbuilding was a waste of taxpayer dollars as well as a detriment to the animals. Sometimes Herman felt he couldn't win.

Part of this road improvement and public access plan was to build cement boat ramps at each of the oxbow lakes made accessible by these roads. Surprisingly, almost nobody complained about adding the boat ramps. None of these lakes had ever had a decent boat ramp, and the people seemed to truly appreciate the luxury.

Another issue that caused a lot of consternation was how to deal with the camps that were on or near the Pascagoula River on land that had been previously owned by the Pascagoula Hardwood Company but was now owned by the State of Mississippi.

Gardiner Greene had promised both the camp owners and Herman that he would ensure the camp owners would be "taken care of" as part of the sale of the property. To put it simply, that didn't happen. There were no legally binding provisions whatsoever for the camp owners included in the sale. The camp owners were at the mercy of the state.

Ironically, a fair number of the camp owners had been employees of the Pascagoula Hardwood Company or one of its affiliates and/or were friends of members of the four families that had owned the Pascagoula Hardwood Company. This relationship with the hardwood company and/or the owners of the company had given the camp owners a false sense of security. They were surprised and disappointed that Gardiner hadn't "taken care of" them as promised. The camp owners were entirely left to their own devices with no possible legal recourse.

The state now owned the land upon which their camps sat, and the powers in Jackson were in no mood to show any leniency at all. Jackson wanted them gone!

Herman personally knew a lot of these "campers." Many of them had camps very near Herman's new home, and he had known them for a long time. He watched many of their kids grow up enjoying the swamp on weekends, holidays, and vacations. Herman knew many of the campers had invested a lot of money and time into their camps. He also knew many of the camps couldn't simply be moved because they were just too large. Herman felt it was beyond unfair to evict not only the camp owners but their camps as well. As far as he was concerned, unceremoniously forcing them out was simply immoral.

This scenario was what Herman had feared for so many years as it pertained to his own home. If and when the land was sold, there was no guarantee the new landowners would honor or extend the leases. Herman had feared losing his home in this manner, and now his friends faced losing their camps.

He went to battle for the camp owners. And what a battle it was. The state considered the camp owners to be interlopers and had no sympathy for their plight. The public considered the camp owners to be "privileged" and also wanted them gone. The biologists, the foresters, the "tree huggers," and the politicians all wanted them gone. Herman seemed to be the only person on the side of the camp owners.

Ironically, some of the camp owners resented that Herman had managed to buy the land on which his home sat. Some of them actually accused Herman of pulling a fast one over on them while selfishly taking care of himself. Once again, Herman found himself fighting for a group of people while certain of these same people were simultaneously attacking him.

Largely due to Herman's efforts, the camp owners won a reprieve. The edict from Jackson was that the camps could remain but couldn't be improved, sold, given away, or inherited. They also couldn't be repaired if they sustained any major damage. For instance, if a hurricane or tornado tore the roof off the camp or if it sustained severe damage from a fire, the camp would have to be declared a loss and abandoned. Basically, it was a plan that would give the current owners a reprieve from eviction but would ensure the camps would eventually disappear.

Jackson wasn't happy with this reprieve, and neither was the public. Herman continued to intercede with the state on behalf of the camp owners for many years. Eventually, the state had had enough, and with public sentiment on their side, Jackson rescinded the reprieve. The camps had to go!

And not only the camps. The various houseboats that were moored to the bank were also told they had to go. They could moor on the private riverbank but not the riverbank owned by the state. Unfortunately, there was little private land ownership of the bank of the Pascagoula River. Once again, the heavy hand of Jackson came crashing down.

It was the end of an era.

The "Big Swamp," as it was called, was a different story altogether. It presented a different set of opportunities and challenges. The Big Swamp was a large area between Black Creek and the Pascagoula River with no legal overland right-of-way access, and the network of old logging roads wasn't nearly as vast. The Big Swamp seemed, on the surface, to represent a much more wild and primitive area than the rest of the Pascagoula Swamp. The "tree huggers" loved it.

In reality, though, the Big Swamp was hunted and fished pretty heavily. Numerous people had ferried old Jeeps across the river and left them in the Big Swamp. What the Big Swamp actually represented was a somewhat private hunting area for a select group of hunters. When the river and/or creek were not flooded, the only ways to access the vast interior of the Big Swamp were either on foot or in one of the Jeeps that had been left near the bank of the river. The general public had practically no access to the interior of the Big Swamp. It had been this way for as long as anybody could remember.

Along with the Jeeps that were full-time residents of the Big Swamp, several permanent dog pens had been built near where the Jeeps were kept. Since the area was so vast, dog hunting for deer was common. Some of the hunters that had Jeeps and dogs living in the Big Swamp had even been known to charge fees for any hunters who wanted to join their group for a day of hunting.

The state, of course, wanted the resident Jeeps, dogs, and dog pens gone. The problem was that it was hard to determine what belonged to whom. The state couldn't very well send notices to or take legal action against the Jeep or dog owners if they didn't know who they were. Also, if the state eliminated the Jeeps, they would be pretty much eliminating any realistic access to a large portion of the Pascagoula River Swamp. Did it really make sense for the state to buy the land, then make it inaccessible to the hunting public—the hunters who had been hunting that area sometimes for generations? It was a tricky situation.

It was a situation that, like so many others, turned into one where Herman was pitted against the Jackson politicians. While Herman realized only a few people had Jeeps in the Big Swamp, he also realized everyone else had the same opportunity to move and leave a Jeep over there. If there got to be too many Jeeps, then something might need to be done, but that hadn't happened over the last few decades and there was no reason to expect it would now become a problem.

As for the dogs and dog pens, Herman believed it was safer for man and dog and certainly easier for the hunters if they left their dogs in the Big Swamp overnight during hunting season rather than ferry all of them back and forth across the river each night.

In other words, Herman believed the situation with the Jeeps, dogs, and dog pens should be left alone. Status quo. He managed to keep the status quo until he retired. It became another story once Herman was out of the picture.

In order to gain control over the Big Swamp and its hunters, the state had to have access. Herman had earlier made arrangements with a gentleman by the name of Reginald ("Reg") Green to build a road across Reg's land in order to have vehicular access. Reg agreed to state-authorized vehicles only. His land was not going to become a conduit for a steady stream of hunters or log trucks.

The swamp and the river had become one and the same to Herman a long time ago. He had fished and hunted the swamp from a boat since he was a child. When the river was up, there were (and are) many sloughs and drains that allowed access to the swamp directly from the river. Some of these sloughs are quite long and can provide access to vast areas of the swamp during high water levels.

This situation was, and remains, particularly true in the Big Swamp. When the river is up, the only realistic access to this vast and remote area is by using the network of sloughs. The locals had long ago figured this out, but Herman felt the Big Swamp should be made accessible to the general public, and what better way than to open up nature's natural thoroughfares so the general public could have access?

Paddling or drifting downstream in a flooded slough is an excellent way to see the swamp. The wildlife isn't as easily spooked by a drifting boat as it is by a vehicle or a man walking through the swamp. And the peace and solitude that can be found while drifting down a remote, often unnamed slough is hard to describe.

Herman saw the vast network of sloughs connecting the oxbow lakes with each other and with the river as a means to grant the public access to the wonders of the swamp. Access for hunting and fishing, yes, but also access for a myriad of other reasons. Drifting down a slough, for instance, is a great way to slip up on some of the many bird species that live in or migrate through the swamp. It is a great way to get up close and personal with the many species of turtles, frogs, spiders, snakes, and other small animals in the swamp.

The sloughs are also great places to fish for catfish when the river is rising. Herman and his family had been catching fish in the sloughs for as long as he could remember. The sloughs are also sometimes the only realistic way to access some of the more remote oxbow lakes for fishing and duck hunting. In short, the sloughs were, in Herman's mind, a great way to access and enjoy the swamp.

Consequently, Herman set about trying to clean out the slough system and make it easily accessible to the public. He believed the sloughs were a great way to make up for the reduced public accessibility caused by the state closing and gating so many of the old logging roads.

Herman had visions of canoeists using the sloughs to explore the swamp. It would have been easy to plan a canoe trip by launching your canoe, kayak, or even pirogue and planning an exit point in such a way that the entire trip would be downstream. The possibilities were endless.

Of course, there were those who were against this effort by Herman to provide better public access. The "tree huggers," as always, were against any effort to provide public access in any way, shape, or form. Then there were the people who considered opening up the network of sloughs as a way of inviting intruders into areas they considered to be their private domains. Even the local game wardens lamented the fact that opening up the sloughs made it harder for them to monitor all the visitors to the WMA. Again, Herman was reminded that you can't please all the people all the time.

Herman was constantly driven by his desire to make the swamp accessible to the public. The public had ponied up to buy this wonderland, and Herman considered it to be the responsibility of the state to make the river and swamp accessible to the public and to put forth some effort to enhance the experience of the Pascagoula River Swamp. In his mind, the people deserved no less.

Managing the game population of the Pascagoula River Wildlife Management Area was one of Herman's main concerns, particularly the deer herd. Although he wasn't a deer hunter, he knew the size and maintenance of the deer herd were crucial to building and maintaining public support for the WMA. Herman could easily remember when the deer herd was so thin in his youth, and he remembered the effort that had been taken early in his career to enhance the herd. He spent a good portion of his career protecting the deer herd in the swamp and was in no mood to see it exploited and decimated at this point. He was in for a struggle.

Herman's immediate "boss" was the regional wildlife biologist, who answered to the state wildlife biologist as well as the powers that be in Jackson. Herman had a distinctly different philosophy related to the deer herd than the biologists. Unfortunately, the biologist was, in fact, Herman's boss and also had the ear of the powers that be in Jackson. This struggle was going to be epic.

Simply put, the philosophical difference between the biologists and Herman was the harvest of doe deer. Herman was of the belief doe deer should never be harvested under any circumstances as long as the swamp could support the herd.

Deer are not monogamous. They are more like cattle when it comes to reproducing. Have you ever seen a rancher who has as many or more bulls than cows? Of course not. One bull can service quite a few cows. The number of cows a rancher has, along with the quality of his bull, is the main determinant of how many calves will be born during calving season.

It's the same with deer. During the rut, a large buck will service numerous does. As with cattle, the number of does in the herd, along with the quality of the bucks, is the main determinant of how many fawns will be born in late summer. It really is that simple. If you "thin" the doe population, you "thin" the deer herd. Period.

Biologists always make the argument that the environment limits the size of the herd. While that is theoretically true, the Pascagoula River Swamp is so lush and diverse that it can provide for a tremendous deer herd. Never in Herman's life had he seen a time when the deer herd was limited by the bounty the Pascagoula River Swamp could provide. Herman could see no reason whatsoever to EVER allow a doe to be legally killed.

The biologists would then argue that while Herman was correct for most of the year, the winter was a different story. Deep into winter, the acorn crop was gone and the natural foraging opportunities were thin. They would argue this yearly period was the limiting factor for how many deer the Pascagoula River Swamp could support. Herman was unconvinced. He agreed the winter period was the hardest on the deer but maintained he had never seen a starving deer in the Pascagoula River Swamp. He wasn't buying their arguments.

However, just to further destroy the biologists' case that the deer herd (meaning does) needed thinning, Herman had a solution—planting food plots, lots of them, all over the swamp. Planting them with winter rye grass, which is greenest during the exact time the biologists said the deer were starving. The deer weren't starving and Herman knew it, but planting lots of food plots made their argument completely moot.

So, Herman planted food plots. Lots of food plots.

And the biologists kept installing hunting regulations that allowed the taking of lots of does. It seemed to Herman that the biologists and politicians were bound and determined to provide ever-growing opportunities to legally harvest does. Herman couldn't win, but he never gave up on this ongoing struggle.

The first few years that he managed the George County portion of the Pascagoula River WMA was a success by most measures. The appropriations were more than adequate as is normally the case with a new project, and despite Herman's differences with the politicians in Jackson, he received considerable cooperation from above. A fair number of people were dissatisfied for an assortment of reasons, but Herman just had to keep telling himself that you can't please all the people all the time.

But a setback was brewing . . . .

In September 1979, the Mississippi Gulf Coast and surrounding areas were hit hard by Hurricane Frederic. Frederic wasn't the strongest storm to have hit the Mississippi Gulf Coast. (At that point in time, the strongest hurricane to have hit the Mississippi Gulf Coast was Hurricane Camille in 1969.) But Frederic was plenty strong, and he pretty much came up the backbone of the Pascagoula River Wildlife Management Area.

Despite the almost direct hit, the Murrah house sustained very little damage, but the devastation was severe and widespread in the swamp. The number of trees that were downed was beyond estimation, and quite a bit of wildlife loss was inevitable.

Since the storm hit in September, the acorn crop wasn't yet developed and now would not have the chance to do so. The annual acorn crop is the lifeblood of numerous species in the swamp. The list of species dependent on the annual acorn crop includes but is not limited to deer, wild turkey, wild hogs, squirrels, wood ducks, and several species of native birds.

Nothing could be done about the acorn crop. Animals would suffer and die over the winter, and there was nothing Herman or anyone else could do about that. This was a depressing time for Herman. His beloved swamp and its inhabitants had taken a tremendous hit. The swamp would, however, survive. Herman knew that but couldn't help but lament the tragic and widespread loss.

Even the fish took a hit. All the leaves on all the trees were blown off during the storm and countless branches, limbs, and whole trees were felled into the oxbow lakes. As all this new vegetation was abruptly deposited, another phenomenon took place: The decomposition of all this dying vegetation that absorbed the oxygen from the water and killed a large portion of the fish.

Despite Herman's depression over the widespread loss of both inhabitants and habitat, there was no time to sit and bemoan the losses. Work had to be done. Lots of work.

Nothing had escaped the devastation. None of the nice roads Herman had built to numerous lakes in the swamp were passable and certainly none of the old logging roads or walking trails. The sloughs were, of course, clogged with debris from the storm. It would take weeks just to assess the damage.

There was nothing Herman could do but roll up his sleeves and go to work. Hunting season would just be a loss. There was no way he could clear all the improved roads, logging roads, trails, and sloughs before hunting season. It would take about a year to get the public access network back up and running. The swamp would eventually heal itself, but these roads, trails, and sloughs had to be cleared.

The relationship between Herman and the powers that be in Jackson remained strained for his entire tenure as Area Manager of the George County portion of the Pascagoula River Wildlife Management Area. If the relationship's strain had to be boiled down to one pervading issue, that issue would be control. Herman believed local control was the most efficient way to run anything, including a wildlife management area. He also believed he was uniquely qualified to exert this local control, given his lifelong intimate relationship with the Pascagoula River Swamp. He fervently believed nobody was better suited than himself to determine what was best for his beloved swamp and naturally bristled at any interference from Jackson. There is little doubt Herman could have and should have prioritized the politics of dealing with Jackson, but he never learned the art of politics.

The people in power in Jackson, of course, were actually a parade of different people over the years. When a new Mississippi governor was elected, they would always appoint a new director to the Mississippi Game & Fish Commission, and the new director would bring in his friends and political cronies and install them into positions of power. Each new group of people had to learn for themselves that Herman Murrah was hard to control. He marched to his own drum.

Herman had what would be considered nowadays an odd sense of right and wrong. He didn't subscribe to the "pick your battles" philosophy. If Herman thought you were wrong, he would tell you so, even if it was a battle he couldn't win. A couple of the battles that Herman pretty much knew he couldn't win but fought anyway were over Sunday hunting and rifles in the swamp.

Herman was always against Sunday hunting, so he fought for years to ban it in the Pascagoula River WMA. He had a religious issue with Sunday hunting and also believed the animals needed a day of rest. He figured that excluding Sunday hunting killed two birds with one stone. He actually prevailed in that battle for a time but eventually lost it.

The swamp is so flat that Herman figured hunting with high-powered, long-range rifles posed too much of a danger. If you missed your target, your errant round could be a life-threatening danger to other unseen hunters for quite a distance. However, so many hunters wanted to hunt with their rifles that Herman didn't prevail in this battle for very long.

Another good example of Herman fighting battles that he knew he couldn't win was his objection to killing doe deer. He knew he was fighting a losing battle, but he never gave up on that war.

Herman fought several battles throughout his entire career. Some of them he won and some he lost, but these relentless battles had a cumulative negative effect on his relationship with Jackson and on his personal health. To put it simply, that little guy in George County was a pain. Being a "pain" would eventually cost Herman his job.

Herman's determination to plant lots of food plots and his willingness to "think outside of the box" came together to create a bizarre situation in one particular instance.

Rather than just planting lots of winter rye patches for the deer, Herman decided to try a few other types of food plots. The first alternative was to plant some chufa plots for the wild turkeys. By this point, he had gotten to know some folks with the National Wild Turkey Federation and they suggested these plots. He planted a few and found the turkeys loved them. Chufa plots became a staple from then on. He didn't want his chufa plots to lead to the mass slaughter of turkeys, so he kept them small and in remote areas and didn't advertise their locations.

Then one day, somebody was complaining to Herman about the deer and rabbits eating their personal turnip patches. Herman had an idea. He decided to plant a turnip patch in the swamp just for the deer, rabbits, and other wildlife. He planted the first big patch of turnips under a power line alongside East Wilkerson Ferry Road on the way to his house. That way, he could easily monitor its progress and use.

Sure enough, the deer, rabbits, and other animals dived right in when the turnips sprouted. There were deer and rabbit tracks throughout the patch of turnips, and it was easy to spot where the animals were eating them.

It seemed he had planted such a large turnip patch that the deer and other animals couldn't keep up with the growth of the turnips in the rich swamp soil, so the turnip patch flourished despite the nightly onslaught by the various wildlife. The next thing Herman knew, he had a successful turnip patch.

Then people found Herman's turnip patch. At night, the animals would eat turnips, and during the day, people would be picking turnips. Technically, the people picking the turnips were breaking the law, but Herman never even considered stopping them. He just planted more turnips.

Some state officials came by one day and found the people out there picking turnips and reminded Herman these people were stealing state property. Herman assured the state officials he would look into it.

Herman did look into it. The turnip patch was starting to look a little thin, so Herman went to town and bought more turnip seeds.

The Pascagoula River snakes through the Pascagoula River Swamp and the Pascagoula River Wildlife Management Area for about fifty miles. As the river flows on this perpetual journey, numerous sandbars are naturally formed. Many of these sandbars are beautiful in their own unique ways. Some are long and flat in the gradual bends of the river, and some are short and high in the sharp bends of the river. Each sandbar offers its own unique opportunities for fishing, swimming, and camping.

Most of these sandbars were accessible via the old logging roads, but most of these old roads had been closed by the state and made off-limits to vehicles. Consequently, the public had almost no vehicular access to any of the sandbars. At Herman's insistence, one sandbar in each county (Jackson and George) was designated as a public sandbar and an improved road was built to that one sandbar. The idea was for that particular sandbar to be accessible to the public even if you didn't have a boat and motor or a Jeep. In George County, that sandbar was the Josephine Sandbar.

The Josephine Sandbar was one of those long and relatively flat sandbars in a gradual bend in the river. The advantage of one of these is the gradual increase in water depth as one leaves the sandbar and enters the river. This gradual water depth makes the sandbars safe places to swim since there are no sudden drop-offs or undertows.

Since the Josephine Sandbar was to be the public's primary interaction point with the Pascagoula River, Herman was determined to make the experience as pleasurable as possible. The Josephine Sandbar was to be the face of the Pascagoula River Wildlife Management Area, and Herman envisioned a place of beauty and serenity where John Q. Public would want to take his family for an afternoon or a weekend.

The first order of business was the road to Josephine. Herman wanted that road to be the best road in the swamp. He envisioned people driving to the sandbar in their new cars without fearing damage to their cars or worrying about getting stuck. He first laid out the road so that it followed the natural ridges as much as possible. Then he brought in dirt as needed to make sure the road didn't form any mudholes or bogs. Herman had a simple philosophy when it came to good roads: the road had to be higher than the surrounding ground so the rainwater would drain from the road. Herman also made sure the road had a considerable crown so it would drain efficiently. Then he added generous amounts of washed gravel. He didn't want anyone to have to drive through a mudhole to get to the Josephine Sandbar.

At the Mound Slough crossing on the road to the Josephine Sandbar, Herman installed what was called a "low-water ford," which occurs when you essentially pave the slough crossing with a steel-reinforced concrete pad to facilitate crossing during times of low water levels.

The low-water ford had several advantages to a bridge. The first and obvious advantage was cost. Building a bridge is obviously expensive. Another advantage of the low-water ford was a reduced impact on the surrounding areas and the environment. In other words, the low-water ford helped preserve the natural beauty of the swamp, and it also allowed the slough to flow naturally. If the slough had been filled in for the crossing, that would have permanently affected the water level in Lower Rhymes Lake from which Mound Slough drained, as well as Upper Rhymes Lake, which drains into Lower Rhymes Lake. Herman wanted to have as little impact as possible on the natural drainage system of the swamp. Lastly, a low-water ford would have no impact on boat traffic on Mound Slough when the river was up. This arrangement fit nicely into his vision of utilizing the natural waterways of the swamp.

Then came the sandbar itself. Herman's vision was a place that was pleasing to the senses and brought the public in direct contact with both the swamp and the river. This was to be a place to bring your kids to swim, much like the public beaches on the Mississippi Gulf Coast. There would be easy access to the sandbar and river. You could set up your umbrella and your beach towels on the sandbar (just like on the Gulf Coast beaches) and watch your kids play in the shallow waters where the Josephine Sandbar meets the Pascagoula River. You wouldn't have to worry about sharks, jellyfish, or stingrays. And you wouldn't have to worry about Highway 90 traffic. In fact, there would be no traffic noise. It would be a place to get away from the hustle and bustle.

It would also be a place to bring your family on an overnight camping trip—or a weeklong camping trip. Herman wanted to keep the camping area as natural as possible. This wasn't going to be an RV park, but he wanted it to be clean and inviting and as close to a natural experience as possible.

The first thing Herman did was to bushhog and clean out a large area at the top of the sandbar for camping. He would make the area open and airy. You would be able to take in the vista of the river and the sandbar and watch your kids play on the sandbar to their hearts' content.

Herman always told people that drinking from the Pascagoula River wouldn't kill you, but it would make your hair fall out (a reference to his bald head). But he knew that drinking from the river wasn't the way to go, so he had an artesian well installed in the camping area so campers would always have ample drinking water.

A walking trail was cut and marked so people could safely and easily commune with nature. The trail meandered through some prime examples of the bottomland hardwood forest for which the Pascagoula River Swamp was famous. The trail also avoided any interaction whatsoever with the public road. It was truly a "wild" trail.

Once Herman was satisfied he had turned the Josephine Sandbar into an inviting place to spend an afternoon or a week with one's family, the next order of business was to keep it that way. Therefore, Herman visited the sandbar often in order to discourage hooligans from interrupting the family experience. He also made sure the road to the sandbar and the camping area on the sandbar were both maintained at all times, especially during the summer swimming and camping seasons. In particular, it was Herman's policy to ensure the entire experience was in top form for the major summer holidays. Prior to Memorial Day, Independence Day, and Labor Day, he would have the road freshly graded, the sides of the road freshly bushhogged, and the camping area in top-notch shape.

Herman considered himself to be a steward of the Josephine Sandbar, and he was determined the people of South Mississippi would have reason to be proud of their heritage.

He believed the area manager was the local "face" of the Mississippi Game & Fish Commission (later renamed "Mississippi Department of Wildlife, Fisheries & Parks"), so he strived to be involved with the community and have them involved with the Pascagoula River Wildlife Management Area.

Herman spearheaded several initiatives to improve the local public access to and involvement with the area manager, the area management, and the area headquarters. These initiatives revolved around the headquarters' facility as well as the area manager. He wanted to make both of these entities publicly accessible physically and personally.

The headquarters needed to be presentable, inviting, and involved in the community so that anyone could drive up to the headquarters at any time of day. If Herman was there, the door was always open. He tried to keep the place clean and neat at all times. The grass was cut often, and everything that was visible to the public was constantly maintained so as to be presentable to the public and exhibit a sense of pride.

In front of the headquarters building, Herman cut a slot into the riverbank and installed a cement boat ramp. This boat ramp served two purposes.

First, it gave Herman and the local game wardens a place to launch their boats in order to work on the river without leaving one of the state trucks at a public boat ramp. This was advantageous for two reasons. Leaving a state truck at a public boat ramp was a security concern and also was a notice to the public that the game warden was on the river. Over Herman's long years as a Game Warden, he had learned it was best if the public was unsure where the Game Warden was at any given time.

Secondly, the boat ramp in front of the office gave the public direct access from the river to the headquarters and the area manager. Anyone who was fishing, hunting, or just riding on the river could pull up to the boat ramp, walk up the hill, and visit with the area manager.

As an added public service, Herman installed a vertical pole alongside the boat ramp and permanently attached a board that measured and displayed the current river level at any time. The markings on the board were synchronized with the official river stage at Merrill.

In order to encourage public involvement with area headquarters, Herman started sponsoring and holding public water safety courses at the building as well as hunter's education classes.

Later on, Herman was able to build a shooting range behind the headquarters building. This was a convenient place for game wardens and other law enforcement officials to practice their pistol marksmanship and even became a place where these LEOs could actually "qualify," saving them all a trip to Jackson. The shooting range was also open to the public. It was used in the hunter's education classes and utilized by quite a few hunters as a place to safely sight in their rifles in preparation for hunting season.

Herman was always looking for a way to make the area headquarters a public place. At the root of pretty much everything Herman did was his firm belief in the concept that the Pascagoula River Swamp belonged to the citizens of the State of Mississippi. While official ownership and title rested with the state, Herman insisted it belonged to the people.

He always made himself available to the public at any time. His personal phone number was published, as well as the phone number of the headquarters. He never ignored a phone call or ever turned anyone away from his door—at headquarters or at home. He believed he was a steward of the Pascagoula River Swamp and a servant of the people.

During his tenure as Area Manager for the Upper Pascagoula River Wildlife Management Area, he finally had the opportunity to carry out a lifelong dream.

Herman had duck hunted a lot while growing up in the Murrah household, where they'd lived off the bounty of the river and swamp. He had gotten quite good at it and enjoyed it immensely. Unfortunately, the duck hunting in George County was paltry compared to the counties in the Delta region of Mississippi. Herman had always dreamed of improving duck hunting in George County.

George County sits in a bit of a dead zone for duck hunting. Migrating ducks tend to travel what have become known as "flyways." The closest flyways to George County were (and remain) the Mississippi River Flyway and the Atlantic Flyway. His local duck hunting grounds fell outside both of these flyways. He had always been jealous of the people living within these flyways and had heard numerous stories of the sky being filled with ducks.

He couldn't do anything about the "flyway" situation, but he had an idea. In the Delta section of Mississippi, numerous "green tree waterfowl areas" had been created, and these areas were a haven for migrating ducks. Herman wanted to build some of these for the local duck hunters.

A "green tree waterfowl area" is a specific type of water enclosure designed specifically to attract and hold migrating ducks. They are called "green tree" because the water is held in the pond or lake during the winter for the ducks but evacuated after duck season so as to not kill the trees—thus the term "green tree." The natural trees that provide cover and food for migrating ducks will die if water remains around their roots throughout the year, so if you just dam up an area and leave the water there year-round, the food-providing trees will die. After a year or so, you wouldn't have any food for the ducks. Conversely, if the area isn't flooded during the migrating season, the ducks won't stop. The water has to be there for one part of the year and cannot be there for another part of the year.

In the Delta, avid duck hunters and wealthy landowners, with the state's aid, solved this dilemma by installing extremely large pumps. They would literally pump the water into the green tree waterfowl area just prior to duck season, then pump the water out of the area after duck season had passed, thus keeping the trees alive or "green."

Herman didn't have sufficient funding to drill the wells and install the pumps that were common in the Delta, so he did the next best thing. He dammed up several areas on the PRWMA for duck season, then literally cut holes in the dams after duck season. In the summer, when the ponds were dry, Herman would plant rice or other crops to provide additional food sources for the ducks. His pride and joy was the Indian Creek Greentree Waterfowl Area at the southern end of the county.

The problem with this arrangement was that he was dependent on the Pascagoula River or local rains to fill the ponds for duck season. Most years, the river, or at least local thunderstorms, would cooperate by supplying the needed water.

Herman's green tree waterfowl areas were successful to a degree. The dearth of migrating ducks outside of the natural flyways remained a limiting factor, but the duck hunting in the Pascagoula River Wildlife Management Area was at least a little better than it had been.

CHAPTER TEN

The Beginning of the End

At first, when the Pascagoula River Wildlife Management Area was created from the land bought from the Pascagoula Hardwood Company, the budget was ample to maintain and improve the area. The PRWMA was a "shiny new toy," so the state didn't mind spending money on it. Herman had plenty of budget to make most of the improvements he wanted.

However, as time passed, the budget allocation started to shrink. The "new and shiny" had worn off. Herman had to start advocating for sufficient funding. He even had to delve into politics a little. It helped when he could specify what project(s) needed continued or additional funding. His job had morphed into a "fundraiser," but he would do whatever he had to do to protect his precious swamp and serve the people who had saved the swamp from clearcutting.

As time passed and the state became even less interested in funding the PRWMA, Herman had to look to other sources of funding. He found there were outside organizations willing to provide additional funding, so once again, he found himself in the position of "salesman."

Two of the main organizations with whom Herman partnered were the National Wild Turkey Federation and Ducks Unlimited.

There were others. However, several problems came along with soliciting funding from these outside sources.

For one thing, they required a specific project to be planned and a proposal submitted for their funding. They wouldn't just give you a bunch of money to do with as you wished. This meant Herman had to spend considerable time applying for grants. This also meant he had to start spending more time than he would have liked at a desk. He much preferred being out in the swamp getting something done rather than sitting in the office. And then, of course, the outside organizations expected a full accounting of how their funds had been spent. Herman's administrative workload went up considerably. He figured he had to do what he had to do.

An additional problem with obtaining funds in this manner reared its ugly head. When Herman applied for and received funding from these outside sources, the state's control over him was diminished. While he had to answer to the outside sources for the funding, he didn't have to answer to the state about how he spent someone else's money. The politicians in Jackson were already unhappy about Herman's penchant for going his own way. Now it was even worse. Their control of Herman was becoming more tenuous with each passing year.

To make matters worse, the outside organizations that were granting funds directly to Herman made corresponding reductions to the funds they would have normally granted to the state. He was competing with the state for funds and was winning. The irritation the state politicians felt toward him was growing exponentially by the year. Power and control are the currencies of the political world, and this guy was "out of control."

During his lifetime, Herman received several accolades and even garnered some national attention. This was primarily due to his role in the state's purchase of the Pascagoula River Swamp from the Pascagoula Hardwood Company.

The first recognition Herman received was of a national nature. *Time* wanted to write an article about the state's purchase of the swamp and wanted a human-interest story for a hook. Herman's personal story and involvement made just such a story. The article was published in the November 8, 1976, issue.

The article accurately portrayed Herman as an integral part of the swamp—an anachronism of sorts. He was pictured in the article admiring a large cypress tree. The article made it clear Herman and the swamp were inextricably linked. They even included his well-known quote about the river: "I drink from it. It'll make your hair fall out, but it won't kill you." This writer has heard Herman utter these words many times. The included photo of Herman showcased his bald head, bringing an ironic clarity to this statement.

Considerable space was devoted in the article to describing the wonder and importance of the Pascagoula River and the Pascagoula River Swamp, including the variety of species of plants, fish, and animals supported by the system. The article even mentioned the endangered species that called the Pascagoula River and its swamp home.

The *Time* article went into substantial detail about how the state had come into possession of the Pascagoula River Swamp, including his involvement as well as that of several of the other key figures and the Nature Conservancy.

Herman was pleased with the final result of the article and so was this writer, who still has an uncirculated copy of this particular issue of the magazine.

Herman was also featured in a book called *Preserving the Pascagoula*, written by Donald G. Schueler and published by University Press of Mississippi in 1980. Herman was prominently featured in chapter 7.

Herman was not overly amused by certain details of the book. He was proud the purchase of the swamp by the state was memorialized in book form by a known author but was less than happy about certain aspects of how the book was written and how the story was told.

According to Schueler, Herman was just an uneducated and clueless anachronist who had almost nothing to do with the whole affair. Nothing could be further from the truth, and nobody knew this better than Herman.

The original idea of the state buying the property was first discussed between Graham Wisner and Herman's son Davy in the summer of 1970. Graham approached Herman with the idea, and Herman was immediately on board. Neither of them believed the idea would ever come to fruition, but the swamp was so important to both of them that they were enthusiastic about trying because they couldn't come up with any other way to save the swamp. The original "lunatic" scheme was hatched by these two men, yet Schueler writes that Herman first heard the idea in 1974 when the process was well underway.

Schueler gives considerable credit for the amazing accomplishment of getting the state to buy the swamp to Avery Wood, then director of the Mississippi Game & Fish Commission. Wood's aide, William C. (Bill) Quisenberry, was also given considerable credit. Herman knew that both these gentlemen were crucial to the whole affair, but Wood would never have known about the idea if Herman hadn't run it up the Mississippi Game & Fish Commission flagpole until it landed on Avery Wood's desk.

Herman was never one to want the spotlight for himself. He wanted the swamp to be saved. Period. He worked hard at making that happen for almost a decade. A lot of people worked hard to make it happen: Avery Wood, Bill "Quiz" Quisenberry, Dave Morine, Charles Deaton, Graham Wisner, and Herman. The dream finally came true. Herman and Graham's dream.

Herman didn't contest the author's version of events. He never complained publicly or even privately. As far as this writer knows, he never said anything about it. It's time to set the record straight. To this day, the Pascagoula River Swamp would probably not exist if it weren't for Herman Murrah and Graham Wisner. These two very different strangers found a common cause in their devotion to the Pascagoula River Swamp, then remained close friends for life.

Herman didn't have time to complain about a book. He had a swamp to improve and make accessible to the people of Mississippi, who ultimately saved the swamp. Had it not been for an upwelling of public support (due in no small part to Quiz), the Legislature would never have approved the funds for the sale, and the swamp would now be one big cottonwood plantation. Herman never forgot the citizens of Mississippi who made his impossible dream come true. In his mind, they were the true heroes of the story.

A little after Herman's retirement from the Area Manager position, the Mississippi Public Broadcasting Network made a documentary about the Pascagoula River and its swamp. It was made in conjunction with the Nature Conservancy and was partially funded by several corporations that had a presence in South Mississippi, including Chevron/Texaco and Mississippi Power.

The documentary was called *The Singing River . . . Rhythms of Nature*, and this writer highly recommends you give it a watch. The narration was provided by Gerald McRaney, a Mississippi native and a well-known and successful actor.

The Pascagoula River is far and away the star of the film. The writers did justice to the river and its uniqueness among rivers in North America. As of this writing, the Pascagoula River is still the largest unobstructed river in the United States. By "unobstructed," the film explains that the Pascagoula River has no dams, locks, or levees and has never been dredged. Leaving the river natural like this allows it to flood into the swamp as it always has. This annual flooding makes not only the Pascagoula River but also the Pascagoula River Swamp unique among its peers. Nowhere else in America does a river have such an intimate relationship with a surrounding swamp.

Speaking of the word "swamp," the film explains the land adjoining the Pascagoula River is not what one imagines when the word "swamp" is used. The Pascagoula River Swamp is nothing like the Everglades. When the river is within its banks, the swamp is solid ground. The more proper term would be "bottomland hardwood forest," but "swamp" rolls more easily off the tongue.

The film's writers do a commendable job of documenting the Pascagoula River and Swamp's importance to migrating birds. It also points out the cultural, archeological, and historical significance of the Pascagoula River Swamp. The film even tells the story of how the "Singing River" got its name based on the legend of the entire Pascagoula Indian tribe committing mass suicide rather than being captured by the Biloxi Indians. They reportedly walked into the river holding hands and singing. Legend has it you can still sometimes hear them singing on still evenings just before sunset.

The film tells an abbreviated version of how Herman, Graham Wisner, and others managed to save the swamp from being sold and clearcut. If there's a human "hero" in the film, that would be Graham Wisner, and he is certainly deserving. He is prominently featured in the film.

Herman is also featured prominently. By the time the film was shot, he was obviously in poor health. He even got in the last word, but unfortunately, he passed on before the documentary was released. The film premiered at the Audubon Center in Moss Point, Mississippi. Numerous dignitaries, along with Herman's family, attended.

Overall, the film is well worth watching. It tells the story of the Pascagoula River and its Swamp very well and gives Herman his due credit for being involved in the life of the river during a crucial time.

Herman had worked diligently to take care of the Pascagoula River Swamp and ruffled a lot of feathers along the way, especially in Jackson. He was nearing retirement age, and the folks in Jackson had had about enough of the perennial thorn in their sides called Herman Murrah. As far as they were concerned, it was time he retired and rode off into the sunset. There was only one small problem with this plan—he had no intention of retiring. He had no health issues to speak of and still had the energy, drive, and determination to manage the Pascagoula River Swamp. Herman was loving life.

At first, the powers in Jackson hinted to him that it was about time to hang up his spurs. When that didn't work, they became insistent. That didn't work, either. Herman had no intention of going anywhere, and the state had no mandatory retirement age. It became apparent that he was going to have to be forced out. They would have to make the case that he physically could no longer handle the job.

Back then, an area manager was also a game warden. The game warden part of the job description entailed annual qualifying with your pistol. When Herman started in this role, the standard issue pistol was a .38 revolver. Years later, the Game & Fish Commission changed the standard issue pistol to a .357 Magnum. The .357 was cumbersome when working in the woods, so many of the seasoned game wardens complained, and the state decided they could continue to carry their .38s. They were also allowed to use their .38s for annual pistol qualifying. About the time they were trying to get rid of Herman was when they (all of a sudden and coincidentally) disallowed qualifying with a .38. This caused an unnecessary hardship on the older game wardens, but Herman managed with his .357 Magnum revolver. What they forgot was that Herman had been an excellent pistoleer all his life. Their strategy didn't work.

So, they tried a different tactic. Once again, all of a sudden and coincidentally, it became a state requirement that all game wardens complete a battery of physical fitness testing on an annual basis. Normally, when an organization installs new and additional physical requirements for a job description, the current employees are "grandfathered in." Not so in this case.

Herman knew he couldn't pass these physical standards. His inability to meet them had, after all, been the whole point. He had to retire. The state had given him no choice. It was the end of his career and the end of an era for the Pascagoula River Swamp and the people of George County.

CHAPTER ELEVEN

Put Out to Pasture

After being forced to retire, Herman didn't handle his retirement very well. In essence, he lost the will to live. He had spent most of his adult life on a mission to protect the Pascagoula River Swamp for generations to come and was officially told his services were no longer needed—or even wanted, for that matter. He felt like life had no purpose anymore. This phenomenon is all too common among people forced to retire, and some believe that life expectancy after a forced retirement is pretty short.

Generally speaking, when people face "forced retirement," it is due to a struggling business or budget shortfalls. In other words, through no fault of their own. Said another way, it isn't personal. In Herman's case, he was convinced it was personal. He had no doubt he was personally forced out and took it as an indictment of his worth. It was personal rejection, and personal rejection is hard on a person—just ask a divorcee who was personally rejected after a long marriage. In many cases, personal rejection can be far more debilitating than a death in one's immediate family.

Gertrude's health had been slowly deteriorating for a number of years, but Herman had remained robust with no apparent health issues right up to his forced retirement. Almost immediately, his health started failing him.

The first sign of this rapid decline was when his "nerve" problem that he'd dealt with for so long turned into full-blown depression. He almost never left the house anymore. He even stopped spending time in the swamp. Depression can be devastating, and it certainly was to Herman.

The "normal" signs of old age showed up almost overnight. Herman started having heart problems, high blood pressure, circulatory problems, respiratory problems, arthritis, etc. You name it, and if it was a symptom of old age, it fell upon Herman like a plague.

The first serious issue happened within a year of his retirement. One of his carotid arteries was found to be blocked. He was fortunate that he didn't have a stroke. The doctors decided to do emergency surgery to clear the blockage. This marked the first time he remembered being in the hospital due to an illness (he had spent a little time in the hospital for a couple of injuries over the years). It was the beginning of a pattern.

Not long after that episode, Herman's heart decided to go out on him. He started having all sorts of heart issues. One Sunday, Davy and his wife found Herman at home alone having a heart attack. The signs were clear, but Herman had apparently decided to just tough it out and call it a life. Davy and his wife loaded him up (almost against his will) and rushed him to the emergency room in Lucedale. He was then rushed to a mobile hospital for emergency open-heart surgery. Multiple bypasses were done. Herman pulled through the heart surgery but was never the same afterward due to a side effect of the surgery: a complete collapse of function in his kidney (not "kidneys"). As it turns out, Herman only had one kidney all along, and it was no longer working. This meant he had to start dialysis treatments and would have to continue them for the remainder of his life.

Dialysis treatments had come a long way since their inception a few decades earlier. However, the protocols, procedures, and equipment were not up to today's standards. These days, home dialysis units are available and can be used when needed. Herman had to report to the dialysis treatment facility in Lucedale. Space and equipment were limited, so the treatments were not optimal in frequency. His health and well-being deteriorated between treatments. Sometimes, he went in for an emergency treatment.

During this time period, he was still recovering from heart surgery, his depression was worsening, and now he was having to deal with dialysis treatments. It was a rough time for Herman and his family.

While Herman's health was going downhill, the Celebrate the Pascagoula event was taking place at the George County Area Headquarters. This event celebrated the twenty-fifth anniversary of the state purchasing the Pascagoula River Swamp. Numerous dignitaries from all over the state attended, including Bill Quisenberry, Graham Wisner, and Charles Deaton. More than three hundred people were present. Herman visited one last time with some of the people he had teamed up with to save the Pascagoula River Swamp.

It was also about this time the documentary *Singing River . . . Rhythms of Nature* was shot. At least this was when the footage that included Herman was shot. His failing health was quite obvious at this point, and it shows in the documentary. He only had months to live.

A few months later, Davy drove Herman to the hospital for the last time and slowed down at Herman's request while driving through the swamp so Herman could have one last look. He only lasted a couple of days in the hospital. He died in a morphine-induced sleep.

The news of Herman's impending demise had reached Jackson. Bill Quisenberry, Charles Deaton, and a couple of other officials drove down to Pascagoula to visit one last time with Herman but arrived at the hospital shortly after he'd passed. The hospital hadn't yet removed him from his bed, so they sat with him for a few minutes. Quiz said a prayer.

Herman's funeral proceedings were, in some ways, typical of a Southern funeral. First, there was the wake attended by a wide range of family, friends, professional acquaintances, and dignitaries. This writer was so distraught at the time that he didn't remember who attended, only that there were a lot of people. The same could be said for the funeral itself. All I can remember is that the funeral service was standing room only with an overflow crowd outside.

The funeral procession from the funeral home in Lucedale to the burial site in Buzzard Roost was another matter altogether. The roads were lined with law enforcement people along the entire route, primarily game wardens who came from all over the state. The tribute was quite moving, as was the twenty-one-gun salute at the burial site.

Herman's body was laid to rest in the Calvary Missionary Baptist Church Cemetery (Gertrude's church), but his spirit lives on in the hearts of his two sons, the collective hearts of the people of South Mississippi, and in the Pascagoula River Swamp itself. This writer feels his presence every time he sets foot in Herman's beloved swamp.

EPILOGUE

My name is Davy Murrah and I wrote this tribute to Herman Murrah, my dad. I am the elder of his two sons. It has been a struggle to call him "Herman" as I wrote these passages. I respected my dad and would never have called him "Herman," but I wanted his name used in the book because he so richly deserves the recognition. It has been a particularly tough struggle to write the last couple of chapters because it has required me to relive those trying times of Dad's last days. My dad was my best friend and losing him has been devastating beyond belief to me.

I'm doing something that is extremely rare from a literary standpoint. Up to this point, I have been writing in the third person and am now switching to first person. From here on out, I will be referring to Herman Murrah as "Dad." He was a great dad and so much more.

Dad, along with a few other critical people, started something great in South Mississippi, and I intend to chronicle the relevant events that have transpired since his death in 2002, about twenty years ago, and bring you, dear reader, up to date as of this writing.

The Herman Murrah Preserve

In 2000–01, the Nature Conservancy tried to buy a piece of property at the north end of George County on the west side of the

Pascagoula River. Unfortunately, they were unable to raise enough money to meet the asking price. This particular piece of property consisted of approximately 1,662 acres and adjoined the north end of the Pascagoula Wildlife Management Area. It would have been a nice addition to the PRWMA.

Since the Nature Conservancy couldn't raise enough money to buy the property, the landowners harvested all the marketable and accessible timber. The land had been devalued quite a bit so the Nature Conservancy could afford to buy it. Unfortunately, the land was now pretty much a wasteland. It will take generations before the land regains its previous majesty.

The Nature Conservancy decided to call the new purchase the "Herman Murrah Preserve" in honor of Dad's involvement in the now historic purchase of the Pascagoula River Swamp by the State of Mississippi and to honor his lifelong dedication to conservation. The rest of the Murrah family and I greatly appreciated having our family name so honored.

The dedication of the property was held on September 27, 2002. Several dignitaries were present along with community leaders, representatives from the Nature Conservancy, and the Murrah family. Speeches were given regarding the Nature Conservancy's role in this purchase and their continued support for land conservation. A biologist gave a speech about the biological importance of the land, and I gave a speech about Herman Murrah, the man, as follows:

> Good morning, ladies and gentlemen. My name is Davy Murrah and Herman Murrah was my dad.
>
> I'm so honored and proud to see all of you here for the dedication of the Herman Murrah Preserve. Some of the other speakers have discussed the importance of preserving our natural heritage and the specific importance of this tract of land. I've been asked to talk a little about Herman Murrah, the man.

As I said, Herman Murrah was my dad. I've heard it said that any male can father a child, but it takes a special person to be a dad. I agree, and Herman Murrah was that kind of special person.

Certainly, one of the most important roles of a dad (and the same applies to a mom) is to teach right from wrong and to instill moral values. I am a firm believer that these teachings can be effectively passed along in only one way—by example. And my dad taught me many life lessons—by example.

I was visiting with him a few weeks before he passed on (after he was quite feeble) and he shook his fist at me in jest. I told him that he didn't scare me and never had. He replied that he never had wanted me to be afraid of him, then proceeded to give me a hug. I was never really afraid of him. I respected him but didn't live in fear of his wrath. My dad didn't have a temper. That's not to say he didn't feel strongly about certain things. In fact, he was passionate about many things. He just didn't have a temper.

My dad whipped me once—and only once. He didn't rule by fear. In fact, he didn't "rule" at all. He was more of a leader and he led by example.

I was thinking the other day and trying to remember any lectures he gave me. You know, I can't even remember his giving me the first lecture.

Dad taught me to respect my elders—by example. He always was respectful to his elders. The terms "Mister," "Mrs.," "Yessir," "Yes, ma'am," "No, sir," "No, ma'am," and other terms of respect became part of my normal vocabulary because I heard them from him and from my mom. We even used "Aunt" and "Uncle" for older members of our community, even when they weren't related to us. By the way, that one whipping that I got was for speaking to my mom in a disrespectful way.

Dad taught me to respect women—by example. He never raised his hand to my mom, nor even threatened to, and I've never raised my hand to any woman, nor even threatened to. But respect for women is much more than that. Dad believed that women are our helpmates and, conversely, we are their helpmates. In other words, a marriage is a partnership. Each person has his/her role in the partnership, but it is a partnership with no tyranny involved, only love and kindness.

Dad taught me how to handle myself in the business world—by example. He taught me that my integrity and my reputation were crucial. He taught me to be honest in all my business dealings and to give a man a day's work for a day's pay. He taught me to appreciate an opportunity when it comes along and to act upon it.

One of the things that Dad taught me that has helped me the most in the business world is to have faith in myself. Dad believed he could do anything he set his mind to, and he believed that I could do anything I set my mind to.

In 1996, I decided to quit a very good job and start my own business. At the time, I had a big house note and a kid in college. It was a risky thing to do. I talked it over with a lot of my friends. Most of the responses were positive. Most of my friends tried to be supportive. They tried to make sure I had thought the thing through and that I hadn't forgotten any detail that might bite me. Nobody tried to talk me out of it, and I was really proud of that. I felt that my friends were being really supportive.

Then I told my dad about it. I remember the conversation like it was yesterday. I told him what I was planning to do and he replied, "It's a good move."

"But Dad, I haven't told you any of the details about just how I'm going to make this work."

I'll never forget what he said: "I don't need to know the details. I know you. You wouldn't be making this move if you hadn't considered all the details yourself. You know you will succeed and so do I." My dad was the only friend who had an unqualified faith in me.

My dad taught me respect and love for the environment—by example. Never did he litter. I've never seen him throw a Coke bottle out the window. When he ate his Vienna sausage sandwich, he put the wrapper in his pocket and took it home. He taught me respect for animals. One rule in our house was that you never killed anything that you weren't going to eat.

When I was a kid, I went squirrel hunting for a little while just before dark one afternoon. I shot two squirrels, but one fell in the middle of a big briar patch. I decided it wasn't worth it to go get him. I came home at about dark, and Mom had supper ready. After supper, I went out back to clean that one squirrel.

Dad came out and asked me, "By the way, where is the other squirrel?" (He had heard the shots. We weren't wealthy people, and we didn't waste shotgun shells. And it's not hard to kill a squirrel with a shotgun.) I told him I only brought one home.

"But you shot twice."

"The other one fell in a big briar patch."

"Was he dead?" (Any of you who have squirrel hunted very much know that you can tell, as a squirrel is falling, whether you killed him or just knocked him out of the tree.)

"Yessir."

"Was that briar patch there when you shot him?"

"Yessir."

"Then I guess you're going back to get him." I then started out the front door. (The woods were out back.)

"Where you going?"

"To get the flashlight out of your truck."

"It wasn't dark when you shot him. You don't need a flashlight."

The next thing I knew, I found myself in that briar patch, after dark, on my hands and knees, literally feeling for that squirrel. I found him—and learned a life lesson that stuck with me the rest of my life. Wildlife isn't there for target practice.

The loss of my dad has been devastating to our family—particularly my mom. Some people may look at this as our being too emotional and maybe weak. I can tell you that my mom is one of the toughest women I've ever known. The loss has devastated her so much BECAUSE it was such a huge loss. We keep reminding ourselves and each other that it hurts so much only because we were so blessed to have such a great dad and husband. We should be, and are, thankful for that.

The Pascagoula River system, our local community, and the State of Mississippi have also suffered a great loss, but it is only a great loss because the river swamp, the local community, and the state were blessed to have such a great man in the first place.

Naming this preserve in his honor is a fitting tribute to a great man who loved this land and dedicated his life to its preservation.

Thank you all for your indulgence and for being here.

For me, it was a great day spent in honor of Dad. The emotions ran high. I was so proud!!

The Nature Conservancy has diligently tried to hasten the recovery of the property after the clearcutting. They have strived to eradicate invasive species and have planted over 100,000 hardwood seedlings.

The creation of the Herman Murrah Preserve occurred after Dad's passing. I will now start back at the beginning of his story and update you on various aspects of Dad's life and what has transpired since his passing.

The People

Let's start with Dad's boyhood friend, Leon. Dad and Leon remained friends for life and beyond. Let me explain "and beyond." What I mean by that is Leon remained Dad's friend long after Dad passed. If you wanted Leon to revert to his old fighting ways, then just start badmouthing his dear friend Herman.

I have had many conversations with Leon over the years, and he would always reminisce about his and Dad's escapades. Leon's stories fit exactly into Dad's stories of their youth. They truly were best friends. Leon, of course, attended Dad's wake and funeral. As

I've said, I was so distraught that I don't remember much about Dad's funeral proceedings, but I remember Leon. Seeing Leon at the wake was an emotional moment for me. That big, tough man was shedding tears over his lifelong friend. Leon passed a few years later. I miss Leon.

Gertrude Havard became Gertrude Murrah and was my mom.

A lot of this book deals with Dad's love affair with the Pascagoula River and the Pascagoula River Swamp. Have no doubt that Dad loved Mom and Mom loved Dad from the bottom of their respective hearts. They loved each other until she passed away about six years after Dad. Mom never even came close to recovering from Dad's passing. They were true soul mates.

As stated earlier, Dad was determined to be more of a loving father than his father had been. Consequently, I never received much corporal punishment from Dad, even though he was a believer in "spare the rod, spoil the child." To be truthful, Dad spoiled me. He only whipped me once that I can remember, and that was for calling Mom by her given name—"Gertrude." Dad would and did put up with a lot of mischief from me but absolutely would NOT tolerate my disrespecting Mom.

As far as I can tell, my younger brother (Danny) and I are the "old men" of the remaining extended Murrah clan at this point. Dad was a man of small physical stature, but neither Danny nor I have even come close to filling his shoes.

Graham Wisner became a lawyer and worked for the Nature Conservancy for a time. He went on to become successful in international law. Dad and Graham remained lifelong friends. To the best of my knowledge, a framed photo of Dad still graces Graham's desk. Graham is retired and currently lives on the outskirts of Washington, DC. I remain in touch with him to this day.

William "Bill" Quisenberry (or "Quiz," as Dad called him) finished out his career with the Mississippi Game & Fish Commission (later named the Mississippi Department of Wildlife, Fisheries & Parks) and actually retired about the same time as Dad. Quiz is enjoying retirement in Clinton, Mississippi. I talked with Quiz just the other day. He and Dad remained friends for life.

Mom lived in Dad's "house on the river" until her passing. The home remains in the Murrah family. Actually, my wife (Gwen) and I bought the house and now make it our home. We have completely renovated the house, but the spirit of Herman Murrah still lives here. The bank of the Pascagoula River is still occupied by the Murrahs.

Developments

My dad loved and cared for the Pascagoula River Swamp like no one before or since. He devoted his life to preserving and, yes, even improving the swamp. It was his dream the swamp would remain a national treasure and that people would be able to enjoy this wonder of nature. Since his retirement and passing, the swamp has gone through a slow decline as it has withered on the vine of neglect. Whether this decline and neglect are laid at the feet of budget reductions or just plain apathy will be for the reader to decide. It is my dream and aspiration that this story of his life and lifelong endeavors will inspire a renewed effort to return the Pascagoula River Swamp to its former glory.

I will now delineate just a few examples of how Dad's dreams have been allowed to either expire or wither on the vine.

Management

Due to budget cuts, the management staff has been greatly reduced. At this writing, there is no longer an active area manager for the George County section of the Pascagoula River Wildlife Management Area. There is one area manager for the entire PRWMA, and he is headquartered in Jackson County. The PRWMA now consists of about 42,000 acres stretching approximately fifty miles along both sides of the Pascagoula River. This expansive area cannot be properly maintained with a skeleton staff, and the neglect shows in many ways.

The Plan

As soon as the state took over the Pascagoula River Swamp, a committee was appointed in order to develop a ten-year management plan for the Pascagoula River Wildlife Management Area. Dad was on that first committee and any subsequent committees for as long as he remained the Area Manager for the George County section.

The Mississippi Natural Heritage Law of 1978 mandated the completion of a comprehensive long-range plan for the area. The first plan was completed in 1981 and was designed to be reevaluated and updated at ten-year intervals. I know the plan was updated at least once because I am in possession of the second plan developed in 1990.

According to this pattern, updated plans should have been developed and implemented at ten-year intervals. I admittedly don't know if subsequent plans were developed or when/if the updates stopped. I did, however, ask some of the state management what plan is in effect at this time and the response was something to the effect of "as far as I know, there is no plan."

The "Big Swamp"

The "Big Swamp" is a large area situated between the Pascagoula River and Black Creek and is one of the wilder places in the Pascagoula River Swamp. There are no public roads to the Big Swamp, so access to this wild area is from either the river or the creek. Over the generations, local people devised imaginative ways to access this area, including ferrying Jeeps and permanent lake boats into the swamp in order to gain access.

A lot has changed since Dad's retirement. The situation in the Big Swamp had been a thorn in the side of Jackson since the state bought the swamp. Once Dad was no longer in the picture, the state was quick to act.

The Jeeps had to go. The state had never been agreeable to the Jeeps as permanent residents in the Big Swamp. As I understand the current situation, one is allowed to leave a four-wheeler over there as long as it is registered with the state.

There are numerous oxbow lakes in the Big Swamp, and it was common for local fishermen to leave an old boat in the lake. This had been going on for generations since the lakes were hard to access during low water when the bream and white perch fishing was good. Shortly after Dad was pushed out of the picture, the state physically removed all the boats left in the various lakes. One is now allowed to keep a boat in the Big Swamp only if it is left on a trailer at the approved site for four-wheelers.

Access to the Big Swamp has been greatly diminished, resulting in reduced opportunities for people to enjoy this wild space.

Natural Waterways (Sloughs)

I'm not sure when it changed after Dad's departure, but the sloughs are no longer being maintained by the Mississippi Department of Wildlife, Fisheries and Parks and haven't been for years. Dad's marked canoe trails no longer exist. Public access to the swamp has been greatly diminished. Several of the oxbow lakes that used to be accessible via the swamp's natural waterways are no longer accessible from the river. Since some of these same lakes no longer have permanent boats, the lakes have become pretty much unfishable during the summer months. A lot of good fishing opportunities have been lost.

The Deer Herd

Since Dad's passing, the deer herd has continued to dwindle, and its decimation continues as of this writing. In my entire life, I have never seen the deer herd as thin as it is now. I currently live exactly where I lived when I was growing up. In the last six months, I have not even seen one single deer in the WMA on the road to my house. Not one!! When I was growing up here, we had to be careful driving the road or we would have a deer for a hood ornament. Also, as I wander through the swamp, I'm not seeing nearly the number of deer signs that I used to see. The swamp isn't completely devoid of deer, but there is no doubt whatsoever in my mind that the deer population in the swamp is the lowest I've ever seen.

Why? Good question. The first answer that generally comes to most people's minds seems obvious—there are more people deer hunting now simply because there are more people. That would be a good assumption, except it's not accurate. There aren't nearly as many people deer hunting now as there were when I was growing up. I've personally witnessed this decline in hunters in general over the last few decades. The official statistics on the decline in the sale of hunting licenses back up my observations. The undisputed fact is there are fewer deer hunters now.

Another argument that I hear from time to time is it's "those damn dog hunters." While the argument may have some merit (there's little doubt that a pack of dogs is more likely to run a deer out of an area than is a hunter stealthily going to his tree stand), the fact of the matter is there are fewer deer dogs in the swamp than there used to be. So, it's hard to blame the decimation of the deer herd on the dog hunters.

So, what is the problem? What has caused the deer herd in the swamp to dwindle down to its current dire straits? I believe I know the answers. In my opinion, there are three (3) reasons you rarely see a deer in the swamp these days. These reasons are:

1. The swamp is now surrounded by corn piles and food plots.
2. There are now very few food plots in the swamp.
3. People are headlighting more due to reduced law enforcement.

I will now discuss each of these three issues and how they have changed since Dad retired.

Baiting Deer (and Turkeys) I grew up in a home headed by a game warden. Back then, it was illegal to bait deer or turkey. In other words, corn piles weren't allowed in the woods. I can remember one incident in particular when Dad found a corn pile in the woods outside the WMA. He lay in wait for a couple of days until the guy came to put out more corn, then arrested him on the spot. He didn't have to catch him hunting over the corn—just putting the corn out there was of sufficient illegality.

I can remember another instance when Dad caught a federal judge (I remember his name, but he has since passed and I don't want to trash his family name at this point) hunting turkey over corn. Dad arrested him even after he pointed out he was a federal judge and that Dad would regret arresting him.

Quite a few years ago, the state legalized baiting deer and turkey. In other words, corn piles were legalized as long as you didn't hunt over the corn. A few years later, the state legalized hunting over the corn piles as long as your hunting spot (shooting house, etc.) was at least one hundred yards from the corn. Finally, the state just gave in completely. These days, you can put out corn and hunt over said corn without any restrictions whatsoever. Also, I keep talking about "corn piles" when, by now, most of the corn is dispersed by automatic corn feeders. These automatic corn feeders literally ring the dinner bell, and the deer learn to recognize the sound of the dinner bell and even become trained to the frequency of the disbursements.

These unfettered restrictions on baiting deer and turkey on private property (NOT in the Wildlife Management Areas) have led to a profusion of feeding opportunities for deer and turkey on the private lands and hunting clubs surrounding the Pascagoula River Wildlife Management Area. The result of this profusion of corn piles, automatic corn feeders, and food plots is to draw the deer out of the swamp. It's literally a case of "greener pastures." It's happening. And it's legal. And my dad would be fighting this development tooth and nail if he were still around.

Food Plots When Dad was around, he planted food plots. LOTS of food plots. Why is this important? I have heard many experts tell me food plots do no good and actually do considerable harm to the environment for benefits that are marginal at best. Others will make the claim food plots "are for hunters, not for the deer." I even had one expert assert that food plots are just for show.

There is merit to all of these expert opinions, and in a vacuum, they are probably correct. These people are, after all, experts. But the Pascagoula River Wildlife Management Area doesn't exist in a vacuum. As stated above, the PRWMA is currently surrounded by corn piles (or automatic feeders), winter food plots, and salt licks. In other words, the PRWMA is competing with its neighbors for a finite deer herd—and losing.

Reduced Law Enforcement As far back as I can remember, there have been two game wardens assigned to each county in Mississippi. That is still the case. There are currently two game wardens in George County. Nothing has changed, right? Yes, it has. Let me explain.

Back when Dad was the Area Manager for the George County portion of the Pascagoula River Wildlife Management Area, there were additional law enforcement officers (LEOs) assigned to the woods. As always, there were two game wardens assigned to each county. In addition, each Wildlife Management Area (WMA) had a dedicated LEO assigned to the area. There was an LEO assigned to the Upper Pascagoula River WMA, a LEO assigned to the Lower

Pascagoula River WMA, and a LEO assigned to the Ward Bayou WMA. The same scenario applied to the other WMAs in the area, such as the Red Creek WMA and Leaf River WMA.

While the two game wardens assigned to each county helped out in the WMAs, they had help in the WMAs and were able to concentrate a little more in the surrounding areas of the county. The LEO assigned to each WMA could call on help from the county game wardens, but the county game wardens didn't have to do it all. The LEO assigned to each WMA knew the WMA inside and out and spent pretty much all his time in his assigned WMA. Dad was both the Area Manager of the WMA and a Game Warden. Today's area managers have no law enforcement authority.

Also, back then, your local game warden was well known in the community and was easily accessible. In fact, one of the game wardens who worked with Dad (Gary Welford) was elected as county sheriff when he retired from the Mississippi Game & Fish Commission. The game wardens were THAT well known in the community. You bought your hunting and fishing license from your local game warden, and he was your go-to guy for any question you had about hunting or fishing regulations. His phone number was published, and he received calls at any time of the day or night. I know all this for a fact because I grew up in one of those game warden homes. There was a constant flow of hunters and fishers through our house. Because of this involvement in the community, Dad got all the scoop on who was doing what. It gave him an advantage against the poachers.

This reduced law enforcement has resulted, of course, in increased poaching—particularly "headlighting." Headlighting is a well-known way to gain an advantage over a deer. It has been going on forever and will probably go on in the future. One of the big problems with headlighters is they are hunting for meat—not trophies. Consequently, does are considered fair game. A fair amount of does are taken this way.

There's no doubt the reduced law enforcement is taking a toll on the deer herd. The extent of this toll is impossible to measure.

In summary, Dad fought several battles over the course of his career to try to protect the deer herd. He had limited success along with repeated failures. Since his passing, the deer herd deterioration has accelerated.

Green Tree Waterfowl Areas

Dad fulfilled a lifelong dream by building several green tree waterfowl areas to enhance the duck hunting in George County and maintained these waterfowl areas until his retirement. As far as I know, the maintenance of these areas immediately ceased as soon as Dad was no longer in the picture.

Unfortunately, when the maintenance stopped, it was apparently wintertime and at least one of the areas was left dammed up, resulting in the eventual killing of all the oaks and other hardwoods in the area. This particular area is the Indian Creek Greentree Waterfowl Area near the lower end of the George County section of the Pascagoula River Wildlife Management Area. The area has become a really big cypress lake and also one of the largest rookeries in this area. However, it is no longer a "green tree" waterfowl area since all the hardwoods have been killed by excessive water saturation.

Moving Forward

In summary, my dad (Herman Murrah) dedicated his life to making the Pascagoula River Swamp better and more accessible to the people of South Mississippi.

As mentioned earlier, I am the eldest son of Herman Murrah and I have dedicated myself to returning the Pascagoula River Swamp to its former glory achieved during my father's stewardship. I have established a public advocacy Facebook page called "Friends of Pascagoula River WMA," and through this page, I am marshaling a grassroots movement of people in order to persuade the State of Mississippi to rededicate itself to taking care of the swamp as my dad would surely have wanted. This process is in its early stages as of this writing.

Please join me in the effort to preserve and protect the Pascagoula River Swamp for generations to come. This was the lifelong dream of Herman Murrah. Let's keep the Swamp Rat's dream alive.

ABOUT THE AUTHOR

Photo courtesy of the author

Davy Murrah is a lifelong resident of the Pascagoula River Swamp in South Mississippi and is the oldest son of Herman Murrah.

Printed in the United States
by Baker & Taylor Publisher Services